Y0-BSD-901

INTRODUCTION TO NOISE ANALYSIS

R.W. Harris T.J. Ledwidge

Applied physics series

Series editor H.J.Goldsmid

1. **Structure and properties of magnetic materials** D.J.Craik
2. **Solid state plasmas** M.F.Hoyaux
3. **Far-infrared techniques** M.F.Kimmitt
4. **Superconductivity and its applications** J.E.C.Williams
5. **Physics of solid state devices** T.H.Beeforth and H.J.Goldsmid
6. **Hall generators and magnetoresistors** H.H.Wieder
7. **Introduction to noise analysis** R.W.Harris and T.J.Ledwidge

INTRODUCTION TO NOISE ANALYSIS

R.W. Harris T.J. Ledwidge

Pion Limited, 207 Brondesbury Park, London NW2 5JN

ISBN 0 85086 041 5

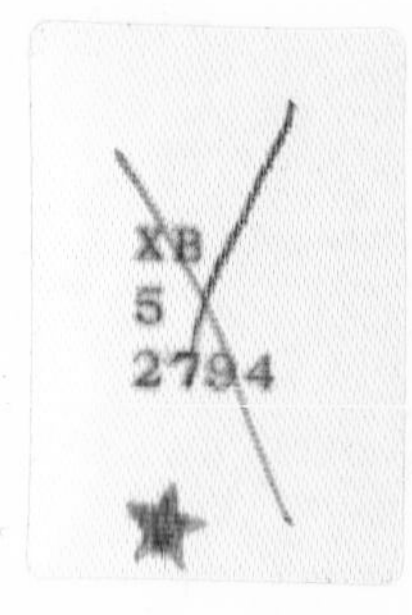

Set on IBM 72 Composers by Pion Limited, London.
Printed in Great Britain by J.W.Arrowsmith Limited, Bristol.

Preface

The subject matter discussed in this book is, in common with most scientific activities, experiencing the effect of the rapid technological expansion in the electronic and computer industry. Mathematical methods that were conceptually known in the time of Lagrange, Euler, and Gauss could not find practical application until the advent of the computer. A second important step in the field of noise analysis was the recent availability of on-line machines which are capable of analysing in very short processing times such important statistical parameters as probability functions, spectral composition, and correlation functions.

We feel that as these machines become generally available to the practising engineers and scientists, the application of noise analysis techniques will increase rapidly. It was with this class of non-specialists in mind that this book was written.

A rigorous approach to the mathematical theory of noise analysis has to a large extent been sacrificed in favour of an approach which tries to highlight the physical understanding of the processes involved in the analysis. It is hoped that this book will serve as an introduction to the many existing excellent texts on the theory of noise and analysis.

In preparing this manuscript we have drawn from a wide variety of sources in addition to our own experience. We have not attempted to compile a comprehensive bibliography, but a suggested reading list is appended. This list contains the titles of the publications which we found most useful and helpful.

Noise analysis is not offered as a universal technique capable of solving all problems, but rather as an adjunct to the classical methods already well known and tried. There are situations in which application of test signals is either impossible or inadvisable and in this case the use of the inherent noise in the system is the only technique.

Broadly speaking the applications of noise analysis techniques fall into three main areas which are:

(i) linear system identification;

(ii) characterisation and recognition of the 'signatures' of physical events;

(iii) measurement of physical parameters.

Some examples taken from these areas are discussed in sections 4 and 5.

In conclusion we hope some of the interest and excitement we have found in this fascinating subject is conveyed by this manuscript to the reader.

R. W. Harris
T. J. Ledwidge

Contents

1

Random signals and their statistical description

1.1 Time and ensemble averages

The existence in nature of fluctuations of a parameter about some mean value is an everyday experience. Temperature measurements taken at equal intervals of time and plotted on a graph provide a simple example of such fluctuating phenomena. A further example would be the height of waves at an instant of time plotted as a function of the distance from the shore.

These fluctuations of physical parameters are commonly referred to as noise. A full and detailed description of the relationship between the coordinates of the noise process is usually far too complicated to be meaningful. We seek, therefore, means to summarise the more important features of such processes in terms which are capable of physical understanding.

The average value of a quantity is a number which provides some useful measure of that quantity. Sometimes the average is taken over a number of samples, for example the average number of peas in a pod; at other times averages taken over a fixed period of time are used.

A distinction is drawn between averages taken over a fixed time or a fixed number of samples. In both cases the limits over which the averages are taken may tend to infinity. The average taken over a given time interval is simply called the **time average**. A collection or set of records describing a process is called an **ensemble** and averages taken across the set are called **ensemble averages**.

Consider the collection of n random signals which are shown graphically in figure 1.1.

Now the collection of records $x_1(t)$, $x_2(t)$, ... $x_n(t)$ constitutes the ensemble, and the ensemble average at a specified time t_1 is given simply as follows:

$$\text{Ensemble average} = \frac{1}{n}[x_1(t_1)+x_2(t_1)+\ldots x_i(t_1)+\ldots x_n(t_1)]$$

or more concisely

$$\widetilde{x(t_1)} = \frac{1}{n}\sum_n x_i(t_1)\,. \tag{1.1}$$

This average is clearly a function of t_1, the time at which the averaging process is specified. It is however conceivable that the same value for the ensemble average arises for all the possible choices of t_1. If this is so the process from which the sample records arose is called a **stationary** one. Mathematically a process is stationary if the following holds for all t

$$\sum_n \frac{x_i(t_1)}{n} = \sum_n \frac{x_i(t_1+t)}{n}\,. \tag{1.2}$$

The time average is defined in the usual way as

$$\overline{x(t)} = \frac{1}{T}\int_0^T x(t)\,\mathrm{d}t\,. \tag{1.3}$$

Note that a solid bar denotes a time average whilst a curly bar denotes an ensemble average.

A further restriction in the class of functions mentioned above occurs if we consider the case when the ensemble average of a stationary process equals the time average. In this case $\widetilde{x(t)} = \overline{x(t)}$ and the process is called **ergodic.**

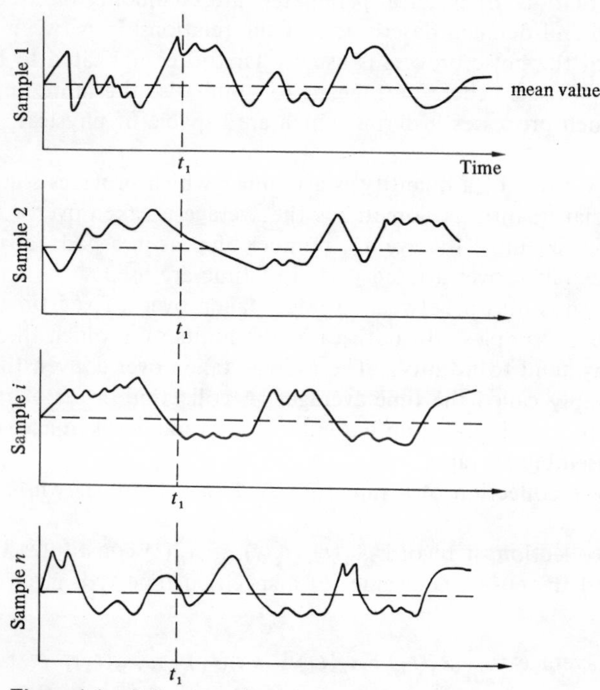

Figure 1.1. Collection of random signals.

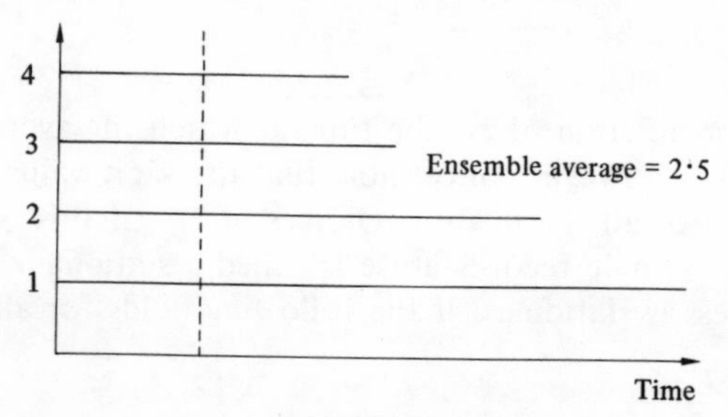

Figure 1.2. Ensemble average of integers.

All ergodic processes are stationary but the converse is not true. As a simple example consider the ensemble to consist of the integers 1 2 3 4. This situation is depicted in figure 1.2.

In this case the ensemble average is 2·5 and does not depend on the choice of origin, and therefore by our definition is stationary. The time average is, however, dependent on the particular member of the set, and the process is therefore not ergodic.

Ergodicity automatically implies that all the statistical properties of the process are invariant in time and that these properties are deducible from measurements made in time.

1.2 Probability distributions

1.2.1 Amplitude probability functions

We have seen that the average value of a signal is one statistical summary that has obvious application. We now examine the probabilistic distribution of the instantaneous value of the amplitude of the signal. The **amplitude probability function** $\mathrm{F}(X)$ is a measure of the probability of the amplitude of a given signal $x(t)$ being less than, or equal to, a specified level X. Now $|x(t)|$ must lie between $-\infty$ and ∞, and clearly the probability that $|x(t)| \leqslant \infty$ is equal to unity (i.e. a dead certainty). Similarly the probability that $x(t) \leqslant -\infty$ is zero. Hence $\mathrm{F}(X)$ must lie between zero and unity. Summarising, we have

$$\mathrm{F}(X) = \text{probability that } x(t) \leqslant X \,,$$

$$0 \leqslant \mathrm{F}(X) \leqslant 1 \,,$$

$$\mathrm{F}(\infty) = 1 \,,$$

$$\mathrm{F}(-\infty) = 0 \,.$$

For an ergodic process $\mathrm{F}(X)$ is the fraction of the total time for which $x(t) \leqslant X$. The general form of the amplitude probability function is shown in figure 1.3, for a signal with zero mean value.

As a specific example consider the case of a sine wave given by $x = A \sin \omega t$ (see figure 1.4).

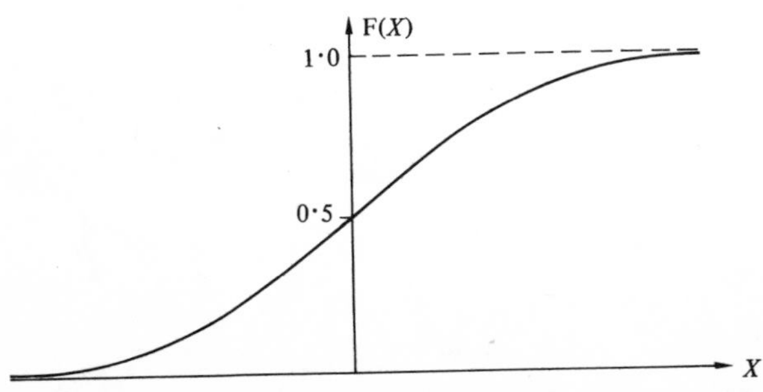

Figure 1.3. General form of amplitude probability function $\mathrm{F}(X)$.

The fraction of time the sine wave spends below the value $x = X$ is given simply by $(2t_x + T)/2T$ where we have

$$t_x = \frac{T}{\pi}\sin^{-1}\left(\frac{X}{A}\right).$$

Substituting this value of t_x we have by definition:

$$\begin{aligned} F(X) &= \frac{1}{2T}\left\{\frac{2T}{\pi}\sin^{-1}\left(\frac{X}{A}\right)+T\right\}, \\ &= \frac{1}{\pi}\sin^{-1}\left(\frac{X}{A}\right)+\frac{1}{2}. \end{aligned} \tag{1.4}$$

This function F(X) is shown graphically below in figure 1.5.

A further example is presented to illustrate this amplitude probability function. Consider a square wave of amplitude $\pm A$. This is shown in figure 1.6 below.

It is clear from the diagram that the probability that $X(t) \leqslant X$ is 0·5 whenever $-A \leqslant X \leqslant +A$, is zero when $x(t) < -A$, and is unity when $x(t) > +A$. The resulting amplitude probability is shown below in figure 1.7.

For an ergodic process the probability of a signal $x(t)$ having an amplitude between X and $X + \delta X$ is simply the probability that

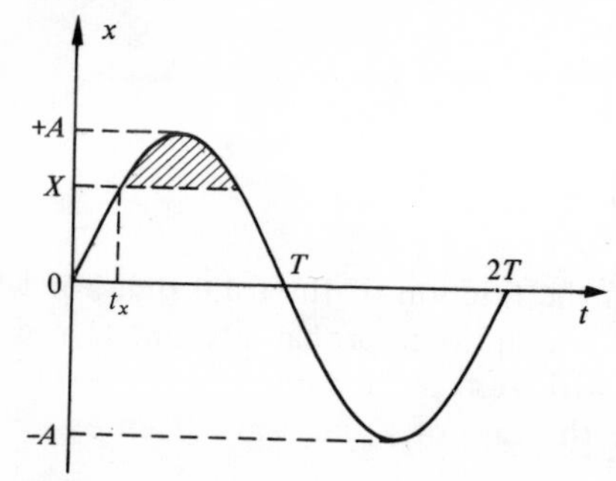

Figure 1.4. Sine wave $x = A \sin \omega t$.

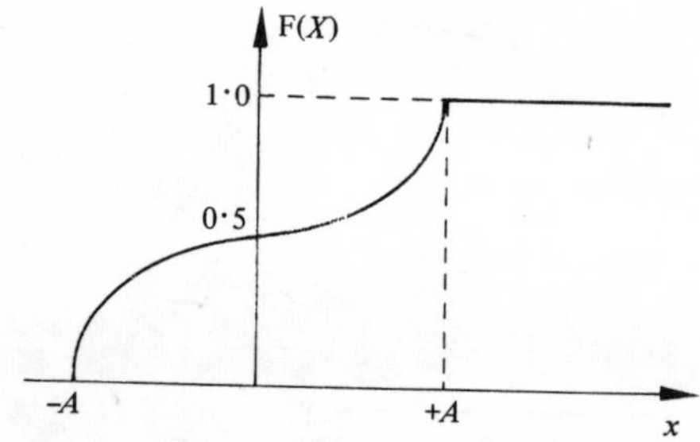

Figure 1.5. Amplitude probability function of a sine wave.

$x(t) \leqslant X+\delta X$ minus the probability that $x(t) \leqslant X$. This may be written as follows

$$\Pr[x(t) \leqslant X+\delta X]-\Pr[x(t) \leqslant X] = \mathrm{F}(X+\delta X)-\mathrm{F}(X)\,, \tag{1.5}$$

$$= \left[\frac{\mathrm{F}(X+\delta X)-\mathrm{F}(X)}{\delta X}\right]\delta X\,. \tag{1.6}$$

The function in the square brackets of equation (1.6) may be recognised as the derivative of $\mathrm{F}(X)$ with respect to X which we shall define as $\mathrm{f}(X)$:

$$\mathrm{f}(X) = \frac{\mathrm{dF}(X)}{\mathrm{d}X}\,. \tag{1.7}$$

$\mathrm{f}(X)$ is known as the amplitude **probability density function** and the probability that $x(t)$ lies within a window δX wide is simply $\mathrm{f}(X)\delta X$. The general properties of the probability density function (p.d.f.) are given below:

$$\mathrm{f}(X) \geqslant 0\,,$$

$$\mathrm{f}(\infty) = 0\,,$$

$$\mathrm{f}(-\infty) = 0\,.$$

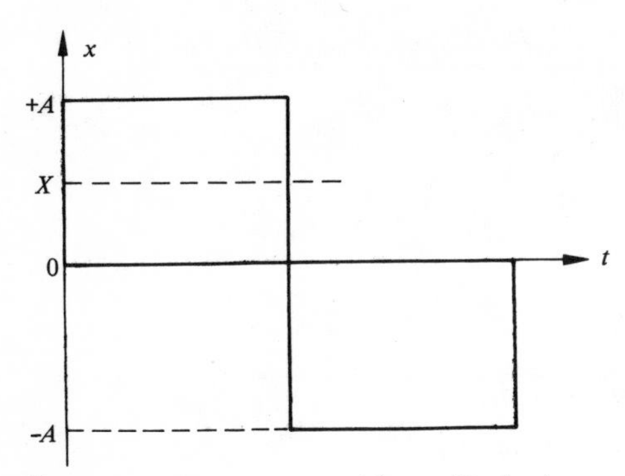

Figure 1.6. Square wave with amplitude A.

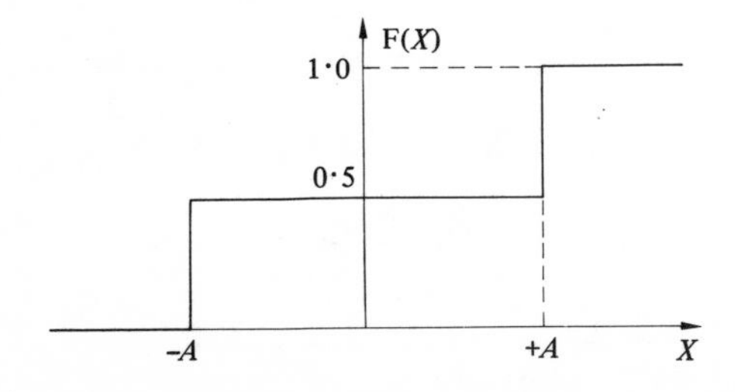

Figure 1.7. Amplitude probability function for a square wave.

Probability that $X_1 \leqslant x(t) \leqslant X_2$ is $\int_{X_1}^{X} f(X)dX$ and in particular $\int_{-\infty}^{\infty} f(X)dX \equiv 1$ i.e. the signal must lie between $+\infty$ and $-\infty$. For an ergodic process it is clear that $f(X)dX$ is simply the fraction of time a signal spends within a given window δX. This idea is illustrated graphically in figure 1.8.

From figure 1.8, we see by definition that

$$f(X)\delta X = \frac{\delta t_1 + \delta t_2 + \dots \delta t_n}{T} .$$

A general form of the p.d.f. is shown in figure 1.9.

The **mode** is defined as the peak of the p.d.f., and the **mean** value $\overline{X}$ has an equal moment of area to the left and to the right of it. This means that

$$\int_{-\infty}^{\infty} (X - \overline{X}) f(X) dX = 0 ,$$

whence

$$\overline{X} = \frac{\int_{-\infty}^{\infty} X f(X) dX}{\int_{-\infty}^{\infty} f(X) dX} , \tag{1.8}$$

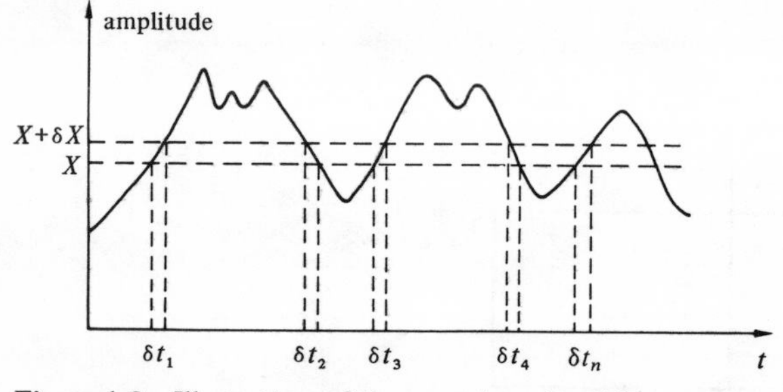

Figure 1.8. Illustration of the meaning of p.d.f.

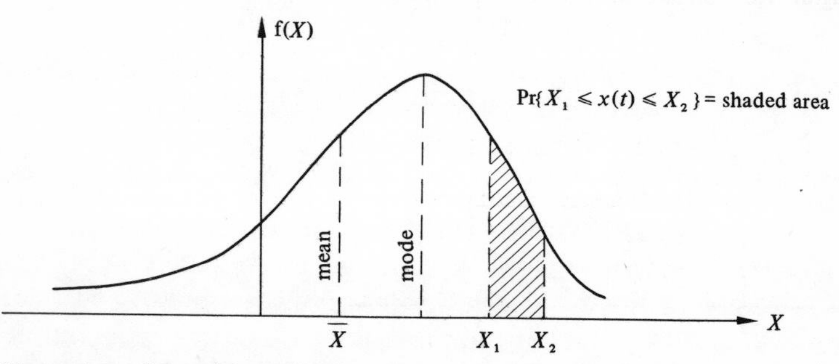

Figure 1.9. General form of the p.d.f.

but as the denominator on the right hand side is equal to unity the mean value is given by

$$\bar{X} = \int_{-\infty}^{\infty} X f(X) \mathrm{d}X \,. \tag{1.9}$$

We see from the above arguments that one way to regard the mean value of a function is to take it as the value on the p.d.f. about which the first moment is zero. We can define the rth moment about the mean μ_r as follows:

$$\mu_r = \int_{-\infty}^{\infty} (X - \bar{X})^r f(X) \mathrm{d}X \,. \tag{1.10}$$

We can without loss of generality consider signals which have zero mean in which case the following results:

1st moment $\mu_1 = 0 = \int_{-\infty}^{\infty} X f(X) \mathrm{d}X$, mean value of signal;

2nd moment $\mu_2 = \int_{-\infty}^{\infty} X^2 f(X) \mathrm{d}X$, mean square of signal or variance of signal.

Clearly higher moments exist and, as we shall see later, are of use in comparing shapes of different probability density functions.

1.2.2 Moment generating functions

A rather interesting way in which higher moments can be generated from a knowledge of the p.d.f. is via the use of **moment generating functions** or **characteristic functions** which are identical in form to Laplace or Fourier transforms discussed later in this book.

The characteristic function $\phi(\alpha)$ is defined as

$$\phi(\alpha) = \int_{-\infty}^{\infty} f(X) e^{-i\alpha X} \mathrm{d}X \,, \tag{1.11}$$

and the rth moment as

$$\mu_r = \frac{1}{(-i)^r} \left[\frac{\mathrm{d}^r \phi(\alpha)}{\mathrm{d}\alpha^r} \right]_{\alpha = 0} . \tag{1.12}$$

One practical use of characteristic functions mathematically transformed from the X plane to the α plane will be in the location of specific non-linearities in system identification.

1.2.3 Theoretical probability density functions

There are several theoretical probability density functions which may be represented by mathematical expressions. If in practice experimental evidence can be fitted by such theoretical distributions to the required degree of accuracy, then further deductions about the data can be made.

The basic distribution discovered in about 1700 by James Bernoulli is the binomial distribution. This distribution arises in the consideration of the relative frequencies of events which are governed by the binomial expression $(p+q)^n$, where p is the probability of success of the event, and $q = 1-p$ is the probability of failure.

The probability of obtaining exactly r successes in n trials in a binomial distribution is simply the rth term in the expansion, i.e.

$$\Pr = \frac{N!}{r!(n-r)!} p^r q^{n-r} . \tag{1.13}$$

As a specific example consider the distribution of the probability of heads in a coin tossing experiment.

In this case $p = q = \frac{1}{2}$, and

$$\Pr = \frac{n'}{r'(n-r)'} \left(\frac{1}{2}\right)^n . \tag{1.13'}$$

Consider the case of $n = 6$, for which we can construct the following table:

Number of heads	0	1	2	3	4	5	6
Probability of occurrence	$1 \times (\frac{1}{2})^6$	$6 \times (\frac{1}{2})^6$	$15 \times (\frac{1}{2})^6$	$30 \times (\frac{1}{2})^6$	$15 \times (\frac{1}{2})^6$	$6 \times (\frac{1}{2})^6$	$1 \times (\frac{1}{2})^6$.

The ordinates of the probability curve of this experiment for discrete values are given by the binomial coefficients. The next logical step is to derive an expression for the envelope of such a binomial distribution. The resulting continuous curve is known as **normal** or **Gaussian** and has the form

$$f(x) = \frac{1}{\sigma(2\pi)^{1/2}} \exp\left[-\frac{(x-\bar{x})^2}{2\sigma^2}\right] , \tag{1.14}$$

where $\bar{x}$ is the mean and σ^2 the variance. Most naturally occurring phenomena have probability density functions which can be approximated by the above expression. It is wrong, however, to assume at the outset that the results of a particular investigation will be Gaussian and it is prudent to test against other theoretical distributions.

The Gaussian distribution as the limit of the binomial is only valid if the probability p is not too small. The **Poisson distribution** arises as the limit of the binomial when the probability p is small. The limits of this theoretical distribution are

$$\left.\begin{array}{l} p \to 0 \\ np \to \text{constant, say } m \end{array}\right\} \text{as } n \to \infty ,$$

and a typical term in the Poisson distribution is

$$\Pr = \frac{m^r e^{-m}}{r!} . \tag{1.15}$$

The Poisson distribution has the interesting property that the mean and variance are both equal to m. Lord Rayleigh introduced yet another variation of the Gaussian distribution when he was studying an acoustic problem. The **Rayleigh distribution** is

$$f(x) = \frac{2X}{\sigma^2} \exp\left(-\frac{X^2}{\sigma^2}\right), \tag{1.16}$$

whereas before σ^2 is the variance. It is apparent that a simple relationship exists between this distribution and the derivative of the Gaussian distribution.

1.2.4 Skewness and kurtosis

The Gaussian curve is usually taken as a reference against which the shape of experimental curves can be compared. Two easily identified parameters of a distribution are the skewness and kurtosis (peakiness) of the curve (see figure 1.10). The Gaussian curve is entirely symmetrical and hence by definition has zero skewness.

Pearson's measure of skewness is

$$\text{skewness} = \frac{\text{mean} - \text{mode}}{\text{standard deviation}}, \tag{1.17}$$

where the mode is the value of the variable which has the maximum probability. This value is easily calculated if the data fit a smooth curve from which the mode can be observed. For an analytical measure it is

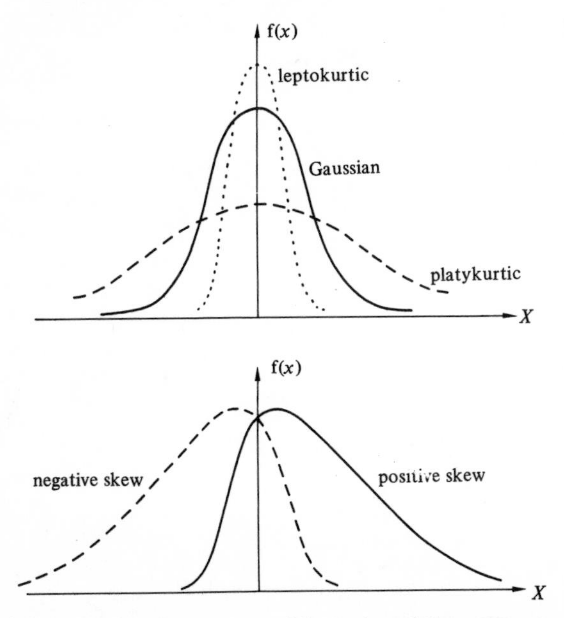

Figure 1.10. Comparison of kurtosis and skewness.

now generally accepted that the following holds:

$$\text{skewness} = \frac{(\beta_2+3)(\beta_1)^{1/2}}{2(5\beta_2-6\beta_1-9)}\,, \tag{1.18}$$

where $\beta_1 = \mu_3^2/\mu_2^3$ and $\beta_2 = \mu_4/\mu_2^2$.

1.2.5 Probability functions for common signals

We will now consider the use of the probability function in calculating some important distributions associated with some simple deterministic and random signals.

Let us start with a sine wave, $x = A\sin\omega t$, shown in figure 1.11. The probability that x lies between X and $X+\delta X$ is

$$\mathrm{f}(X)\delta X = \frac{2\delta t}{T} = \frac{\delta t\omega}{\pi}\,.$$

Now

$$\delta X \approx \frac{\mathrm{d}X}{\mathrm{d}t}\delta t = \omega A\cos\omega t\delta t\,, \tag{1.19}$$

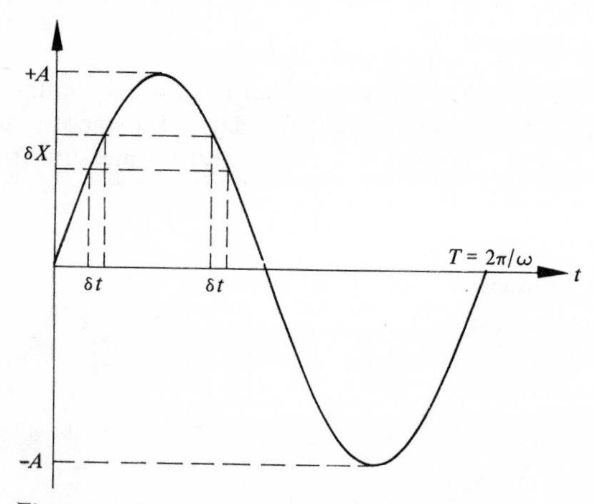

Figure 1.11. Sine wave.

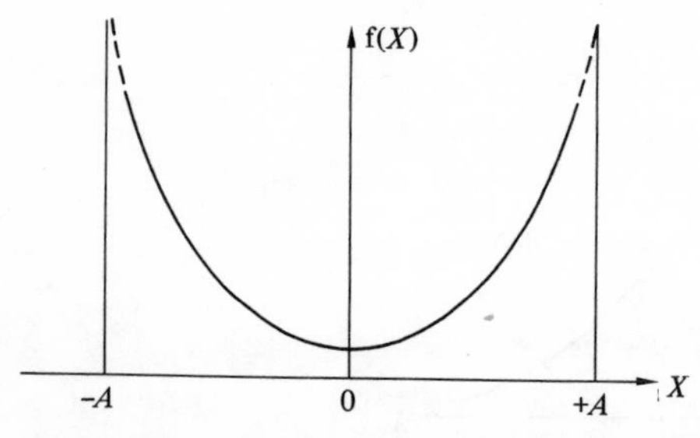

Figure 1.12. Probability density function for a sine wave.

whence

$$\mathrm{f}(X) = \frac{1}{\pi A \cos \omega t} = \frac{1}{\pi A (1 - X^2/A^2)^{½}} \ .$$

This distribution is shown in figure 1.12.

The mean value is

$$\int_{-A}^{A} X\mathrm{f}(X)\mathrm{d}X = \int_{-A}^{A} X \frac{1}{\pi A (1 - X^2/A^2)^{½}} \mathrm{d}X = \frac{1}{\pi A}\left(1 - \frac{X^2}{A^2}\right)\frac{A^2}{2}\bigg|_{-A}^{A} = 0 \ ,$$

as expected. The mean square is

$$\int_{-A}^{A} X^2\mathrm{f}(X)\mathrm{d}X = \int_{-A}^{A} X^2 \frac{\mathrm{d}X}{\pi A (1 - X^2/A^2)^{½}} = \frac{A^2}{2} \ . \tag{1.20}$$

Alternatively we can consider the previous definition of $\mathrm{f}(X)$ as the derivative of $\mathrm{F}(X)$. We previously evaluated $\mathrm{F}(X)$ for a sine wave as

$$\mathrm{F}(X) = \frac{1}{\pi}\sin^{-1}\left(\frac{X}{A}\right) + \frac{1}{2} \ , \tag{1.21}$$

whence, as before,

$$\mathrm{f}(X) = \frac{1}{\pi A} \frac{1}{(1 - X^2/A^2)^{½}} \ . \tag{1.22}$$

As a second rather interesting example consider a signal known as a random triangular wave illustrated in figure 1.13.

As the waveform is made up of straight lines the summation of the small intervals of time δt is a constant, independent of the ordinate X provided $|X| < A$. This means that $\mathrm{f}(X)$ is a constant, whence

$$\int_{-A}^{A} \mathrm{f}(X)\mathrm{d}X = 1 \ ,$$

and

$$\mathrm{f}(X) = \frac{1}{2A} \ . \tag{1.23}$$

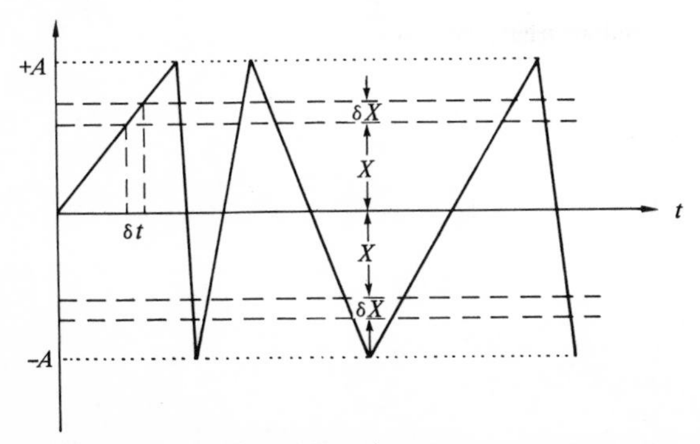

Figure 1.13. Random triangular wave.

This distribution is shown in figure 1.14.

The mean value is

$$\int_{-A}^{A} \frac{X^2}{2A} dX = \frac{A^2}{3} . \tag{1.24}$$

It is interesting at this stage to see the use of the characteristic function applied to this random triangular wave.

The characteristic function is

$$\phi(\alpha) = \int_{-\infty}^{\infty} f(X) e^{-i\alpha X} dX \tag{1.25}$$

and on putting the value of f(X) from equation (1.23) this becomes

$$\phi(\alpha) = \int_{-A}^{A} \frac{1}{2A} e^{-i\alpha X} dX = \frac{1}{2A} \frac{1}{(-i\alpha)} e^{-i\alpha X} \Big|_{-A}^{A} = \frac{\sin 2A}{2A} . \tag{1.26}$$

This is a well-known curve called the sinc function and it is shown in figure 1.15.

We can now use this characteristic function to calculate the moments of the random triangular wave and compare the results to those obtained previously.

The 1st moment, or the mean value, is

$$\frac{1}{(-i)} \left[\frac{d\phi(\alpha)}{d\alpha} \right]_{\alpha = 0} ;$$

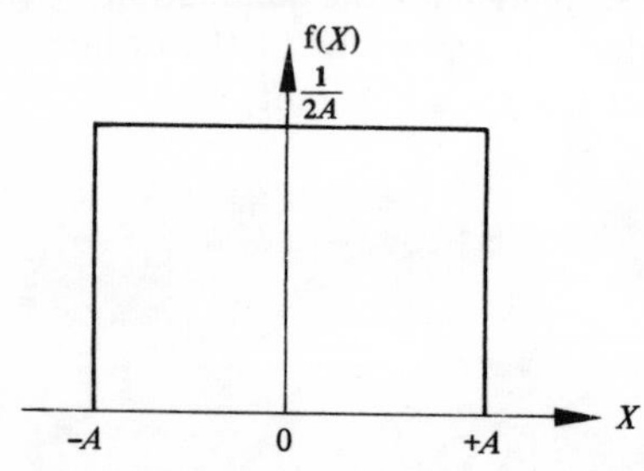

Figure 1.14. p.d.f. of a random triangular wave.

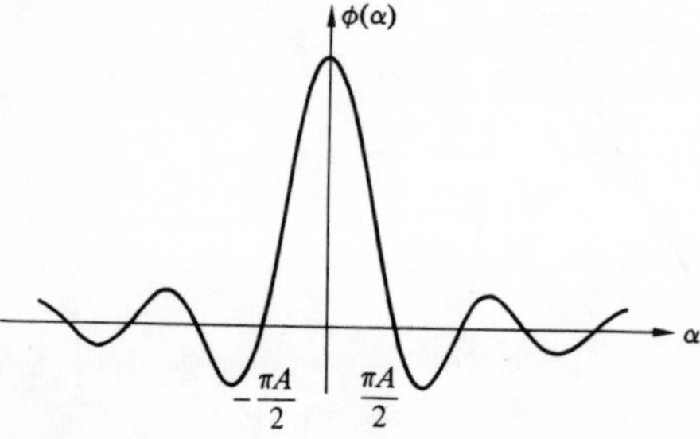

Figure 1.15. Characteristic curve for a random triangular wave.

but

$$\frac{d\phi(\alpha)}{d\alpha} = \frac{\alpha A^2 \cos 2A - A \sin 2A}{\alpha^2 A^2}$$

and it is easily shown that the limit of this expression as $\alpha \to 0$ is 0 whence the mean value = 0 as before.

The 2nd moment, or the variance, is

$$\frac{1}{(-i)^2}\left[\frac{d^2\phi(\alpha)}{d\alpha^2}\right]_{\alpha=0}$$

where

$$\frac{d^2\phi(\alpha)}{d\alpha^2} = \frac{1}{\alpha^4 A^4}\{\alpha^2 A^2[-\alpha A^3 \sin\alpha A + A^2\cos\alpha A - A^2\cos\alpha A]$$
$$- 2\alpha^2[\alpha A^2\cos\alpha A - A\sin\alpha A]\},$$

which can be reduced after some tedious algebra to yield a value for the variance of $\frac{1}{3}A^3$ as before.

The important point to realize is that although the original signal of a random triangular wave defies a precise instantaneous description, the p.d.f. and all its moments give steady unambiguous values. The characteristic function is formally related to the probability density function via the integral transform relationship, and merely expresses the same information in a different way. There is no new information yielded by a transform, but simply a better insight may be gained in looking at the data in a different way.

Figure 1.16 gives a graphical summary of some probability functions.

1.3 Joint distribution functions

1.3.1 Mathematical description

If x_1 and x_2 are two random variables then $F(X_1, X_2)$ is the probability that both $x_1 \leqslant X_1$ and $\leqslant X_2$. By comparison with the single variable case we have the joint p.d.f.,

$$f(X_1, X_2) = \frac{\partial^2 F}{\partial X_1 \partial X_2},$$

which is the probability that both $X_1 \leqslant x_1 \leqslant X_1 + \delta X_1$ and $X_2 \leqslant x_2 \leqslant X_2 + \delta X_2$. Now

$$f_1(X_1) = \int_{-\infty}^{\infty} f(X_1, X_2)\,dX_2,$$

$$f_2(X_2) = \int_{-\infty}^{\infty} f(X_1, X_2)\,dX_1;$$

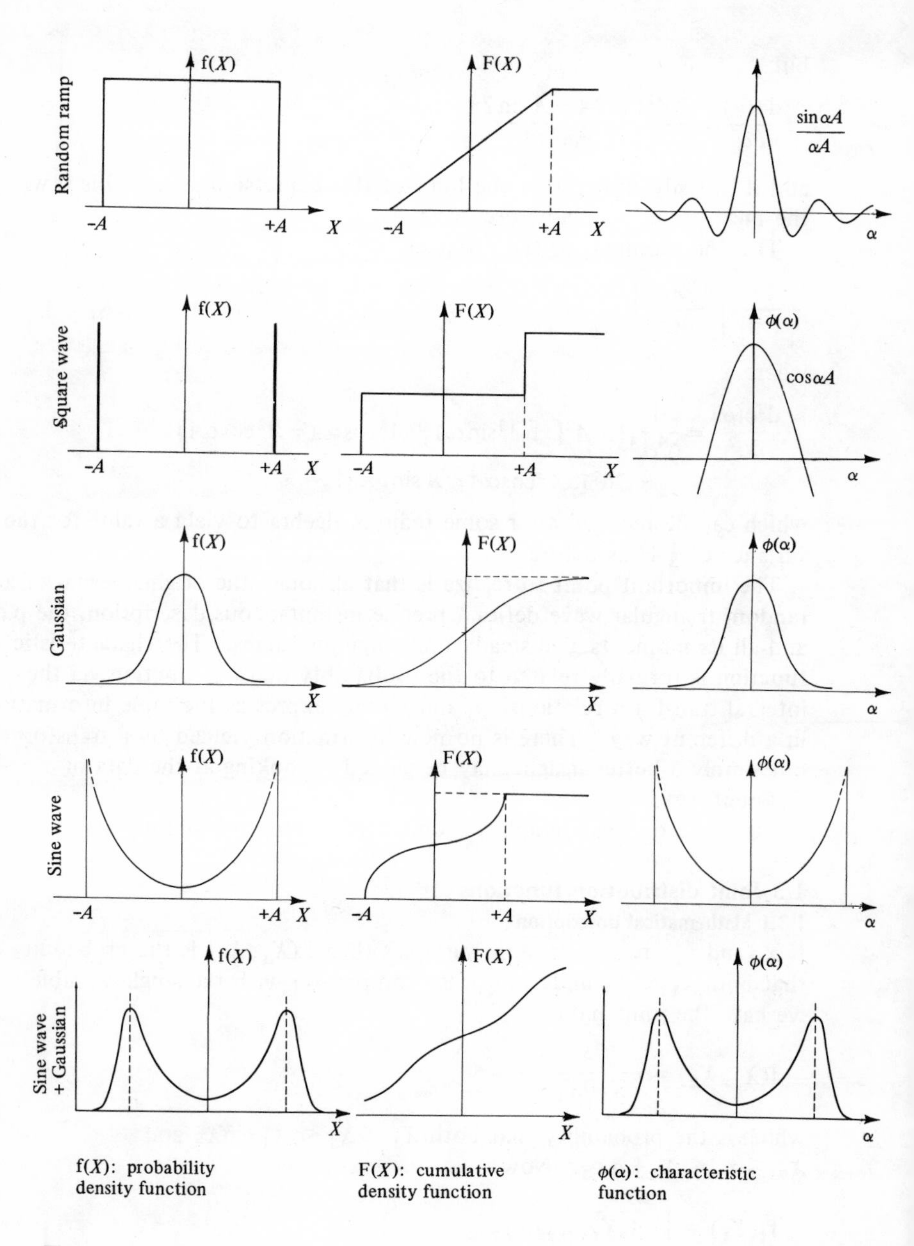

Figure 1.16. Graphical summary of probability densities for various signals.

hence the mean value of x_1^2 is

$$\int_{-\infty}^{\infty}\int_{-\infty}^{\infty} X_1^2 \mathrm{f}(X_1, X_2)\mathrm{d}X_1 \mathrm{d}X_2 \, . \tag{1.27}$$

1.3.2 Graphical interpretation

Figure 1.17 shows a representation of the three-dimensional joint probability. The probability that x_1 lies between X_1 and $X_1 + \delta X_1$ and x_2 lies within X_2 and $X_2 + \delta X_2$ is the volume defined by these ordinates. At the present time there is little practical use made of the joint p.d.f., although it can be used to determine whether two random signals are related, by examining a cross section in the X_1, X_2 plane (circular: no relation; straight line: perfectly related; inclined ellipse: some relationship).

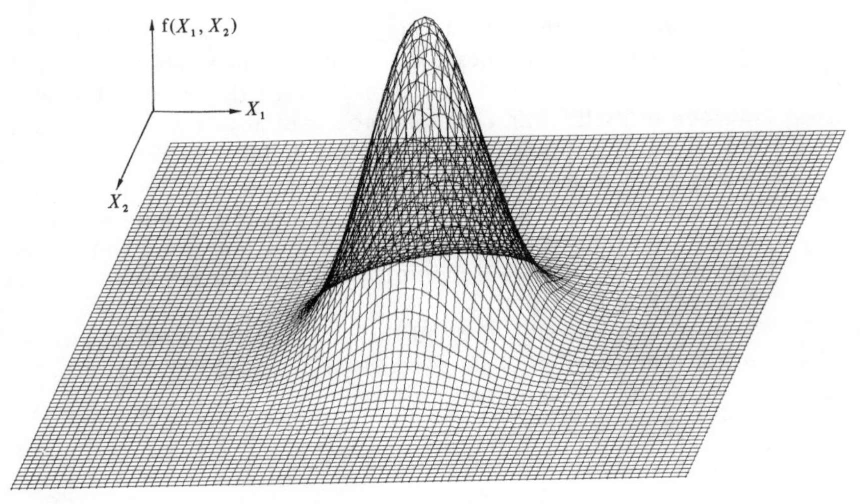

Figure 1.17. Graphical interpretation of joint distribution.

1.4 Stochastic and Markovian processes

Stochastic and Markovian processes are frequently referred to in the literature on random noise applications. The word stochastic is derived from the Greek and implies 'by trial and error' or 'hit and miss'. Common usage now indicates that a stochastic process is a process in which the phenomena are governed by probabilistic laws. This is in contrast to ordered or deterministic phenomena in which each step can be predicted accurately. Typical examples of stochastic processes are:

(i) Radioactive decay of an element. In this case each individual occurrence is random in time though the average rate of occurrence or emission of particles is determined by the element undergoing the decay process.

(ii) Fluctuations of the number of neutrons in a given volumetric space in a nuclear reactor. Here again the average number of neutrons in any given space can be determined from a knowledge of the system. The instantaneous number however is a random variable related to the probability of fissionable occurrences. A Markovian process is a particular form of a stochastic process in which the future state of the process depends only on the present state and not on the route by which the present state was arrived at. A discrete Markovian process is one which proceeds in a series of finite steps each occurring in a random manner with no relationship to the past history of the process.

A Markovian process can be shown to have an autocorrelation function of the form of a decaying exponential function:

$$R_{xx}(\tau) = R_{xx}(0)\mathrm{e}^{-|\tau|/\lambda} , \tag{1.28}$$

where $R_{xx}(0)$ is the correlation of the function with itself and is simply, for a signal with zero mean, the variance. The symbol $|\,|$ denotes a modulus. $R_{xx}(\tau)$ is the correlation between the signal and a delayed version of itself. For a random process of any kind $R_{xx}(\tau)$ must die away to zero as the correlation time tends to infinity. λ is a characteristic decay time or decay length and, if τ is measured in the Lagrangian sense as a distance, λ is in fact identical to G. I. Taylor's definition of characteristic length and finds application in the study of turbulence and similar problems. A fuller discussion of correlation functions and some of their uses is given in chapter 2.

It is now generally accepted that if a stationary random process has a correlation function of the form given by equation (1.28) then the process is taken to be a Markovian process.

Frequency and time description of random signals

2.1 Duality of time and frequency

If we consider the deterministic signal $x(t) = A \sin \omega t$, where $\omega = 2\pi f$ is called the angular frequency, we can specify the value of $x(t)$ at any time by knowing A and ω, i.e. the signal is completely determined for all time by a knowledge of two parameters. The amplitude A is a description of the size of the signal and is analogous to the amplitude probability discussed in chapter 1. It should be realized that the amplitude of a signal tells us nothing about its time history. The converse is also true: if we know the frequency of the deterministic signal given above, we know nothing about its amplitude. It is therefore necessary to have available a statistical description in the frequency domain, and it is natural to think in terms of the average frequency of a random signal, the frequency content, and so on.

Let us return for a moment to a consideration of the deterministic signal. We can say for example that a particular sine wave has a frequency of 1 kHz or a period of 1 m s. Both these statements are equivalent, and it is important to realise that no new information is yielded by transforming from the frequency domain to the time domain; simply the same information is presented in a different form.

As a further example of the duality of time and frequency consider the description of a simple mass–spring–dashpot system shown in figure 2.1.

The relationship between $y(t)$ and $x(t)$ can be easily written down in terms of a differential equation. Clearly the output $y(t)$ will depend on

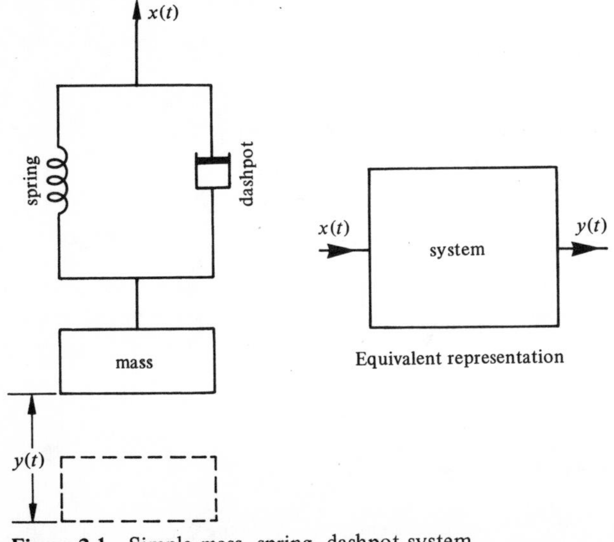

Figure 2.1. Simple mass–spring–dashpot system.

$x(t)$ and the constants; spring constant, mass, and damping coefficient of the system. Various classical forms have been considered for $x(t)$: an impulse, a ramp, a step, and a unit sinusoid.

If $x(t)$ is an impulse, $y(t)$ is the impulse response of the system; if $x(t)$ is a ramp, $y(t)$ is the ramp response of the system; if $x(t)$ is a step, $y(t)$ is the step response of the system; if $x(t)$ is a unit sine wave, $y(t)$ is the sine response.

The first and the last of these functions are interesting because they are capable of being easily related. To illustrate this further let $x(t)$ be an impulse defined as $x(t) = \delta(t)$ where

$$\int_{-\infty}^{\infty} \delta(t)\,\mathrm{d}t \equiv 1 \,. \tag{2.1}$$

$\delta(t)$ is called a delta function and has the property that it is non-zero for only one value of time, as well as having a unity area as expressed by equation (2.1). Then the impulse response $y(t)$ is usually called $h(t)$; for the spring–mass–dashpot system it has the form shown in figure 2.2.

If $x(t)$ was a sine wave then $y(t)$ would, for a linear system, also be a sine wave at the same frequency. The amplitude and phase of $y(t)$ with

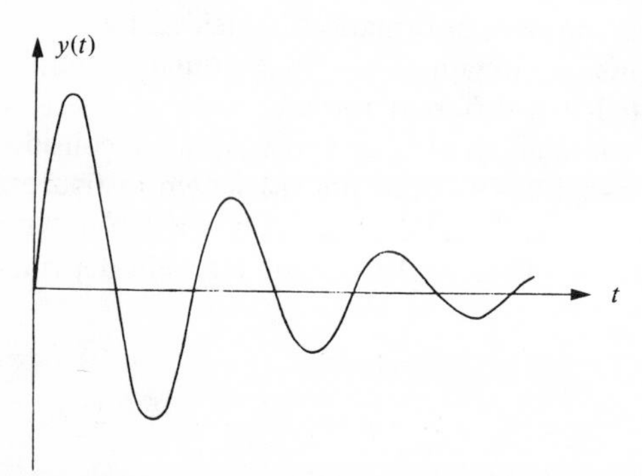

Figure 2.2. Typical impulse response of system shown in figure 2.1.

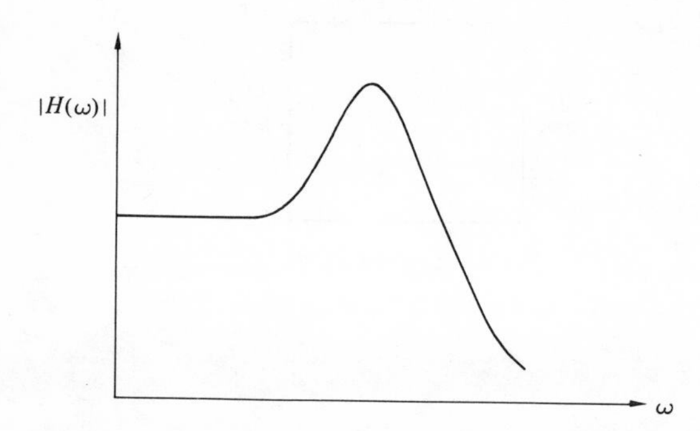

Figure 2.3. Amplitude response of system shown in figure 2.1.

respect to $x(t)$ would obviously depend on the frequency of the input signal $x(t)$ and the nature of the system. The amplitude ratio $|y(t)|/|x(t)|$ and the associated phase angle is usually called the frequency response and written $H(\omega)$. The form of $|H(\omega)|$ is shown in figure 2.3.

Although the shapes of the graphs in figure 2.2 and 2.3 are different they carry information about the same system. The first describes the system in the time domain, the second in the frequency domain. $h(t)$ and $H(\omega)$ are related mathematically by a **Fourier transform** defined as follows:

$$H(\omega) = \int_0^\infty h(t)\mathrm{e}^{-\mathrm{i}\omega t}\,\mathrm{d}t\ , \tag{2.2}$$

$$h(t) = \frac{1}{2\pi}\int_{-\infty}^{\infty} H(\omega)^{\mathrm{i}\omega t}\,\mathrm{d}t\ . \tag{2.3}$$

Note that $H(\omega)$ is complex and contains both amplitude and phase information.

2.2 Analysis in the frequency domain

2.2.1 Fourier transform as a particular integral transform

The integral transform $\mathrm{F}(p)$ of a function of x, $\mathrm{f}(x)$, is defined by the integral equation

$$\mathrm{F}(p) = \int_a^b \mathrm{f}(x)\mathrm{K}(p,x)\,\mathrm{d}x;\ \mathrm{K}(p,x)$$

is a specified function of p, and x is called the **kernel** of the transform.

The effect of applying an integral transform to a partial differential equation is to exclude temporarily a chosen independent variable. The solution of the remaining equation will be a function of p and the remaining variables. This solution has to be 'inverted' to recover the 'excluded' variable.

There are a number of integral transforms in general use in mathematical physics:

$$\mathrm{F}(p) = \int_0^\infty \mathrm{f}(x)\mathrm{e}^{-px}\,\mathrm{d}x\ , \quad \text{Laplace transform;}$$

$$\mathrm{F}(p) = \int_0^\infty \mathrm{f}(x)_{\cos xp}^{\sin xp}\,\mathrm{d}x\ , \quad \text{Fourier } {}_{\cos}^{\sin} \text{ transform;}$$

$$\mathrm{F}(p) = \int_{-\infty}^\infty \mathrm{f}(x)\mathrm{e}^{-\mathrm{i}px}\,\mathrm{d}x\ , \quad \text{complex Fourier transform;}$$

$$\mathrm{F}(p) = \int_0^\infty \mathrm{f}(x)xJ_n(px)\,\mathrm{d}x\ , \quad \text{Hankel transform} \tag{2.4}$$

$[J_n(px)$ is a Bessel function];

$$\mathrm{F}(p) = \int_0^\infty \mathrm{f}(x)x^{p-1}\,\mathrm{d}x\ , \quad \text{Mellin transform.}$$

Note that in a large class of practical problems which start at $t = 0$, Laplace and complex Fourier may be interchanged simply by writing $(\mathrm{i}\omega)$ in place of p in the transformed equations.

2.2.2 Fourier transform as a limit of Fourier series

Consider a function of time which has period T. We shall call this a periodic function, $x_T(t)$ (figure 2.4).

$x_T(t)$ may be written as a Fourier series as

$$x_T(t) = \frac{A_0}{2} + \sum_{r=1}^{\infty} (A_r \cos r\omega_0 t + B_r \sin r\omega_0 t) . \tag{2.5}$$

If we now write

$$\cos r\omega_0 t = \frac{1}{2}(\mathrm{e}^{\mathrm{i}r\omega_0 t} - \mathrm{e}^{-\mathrm{i}r\omega_0 t})$$

$$\sin r\omega_0 t = \frac{1}{2\mathrm{i}}(\mathrm{e}^{\mathrm{i}r\omega_0 t} + \mathrm{e}^{-\mathrm{i}r\omega_0 t}) ,$$

we get the complex form of the Fourier series:

$$x_T(t) = \frac{1}{2T} \sum_{r=-\infty}^{r=\infty} X_r \mathrm{e}^{\mathrm{i}r\omega_0 t} , \tag{2.6}$$

where

$$X_r = \int_{-T}^{T} x_T(t) \mathrm{e}^{-\mathrm{i}r\omega_0 t} \,\mathrm{d}t . \tag{2.7}$$

If we now let $T \to \infty$ we have the description of a non-repetitive function when the summation converges to an integral:

$$X(\mathrm{i}\omega) = \int_{-\infty}^{\infty} x(t) \mathrm{e}^{-\mathrm{i}\omega t} \,\mathrm{d}t ,$$

$$x(t) = \frac{1}{2\pi} \int_{-\infty}^{\infty} X(\mathrm{i}\omega) \mathrm{e}^{\mathrm{i}\omega t} \,\mathrm{d}\omega . \tag{2.8}$$

Fourier transform pair [cf $h(t)$ and $H(\mathrm{i}\omega)$]

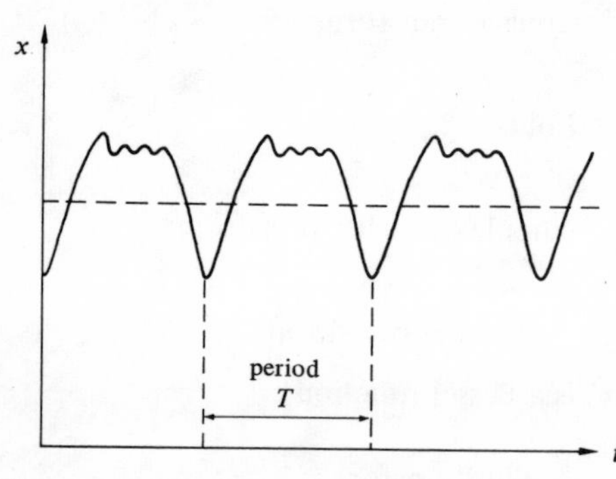

Figure 2.4. Example of periodic function.

2.2.3 Fourier transforms and power spectral density

A sufficient condition for a Fourier transform of $x(t)$ to exist is that $\int_{-\infty}^{\infty} |x(t)|\mathrm{d}t$ is finite. It may not be clear in practical situations whether this condition is satisfied. We can however define a truncated form of the function $x(t)$ which we denote as $x_T(t)$, as shown in figure 2.5. Clearly $x_T(t)$ satisfies the condition. We define the Fourier transform of $x_T(t)$ as follows:

$$X_T(\mathrm{i}\omega) = \int_{-T}^{T} x_T(t)\mathrm{e}^{-\omega t}\mathrm{d}t\,. \tag{2.9}$$

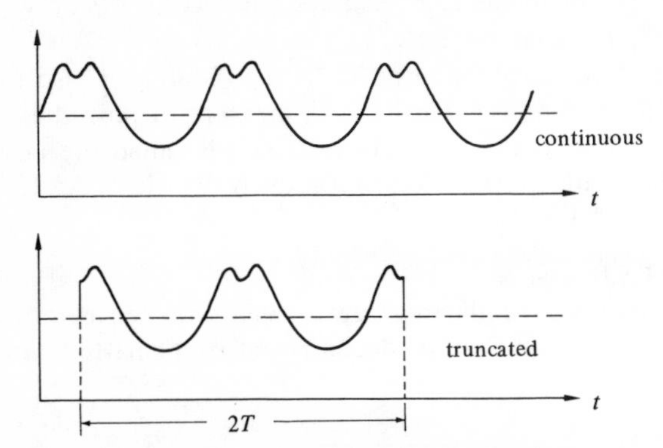

Figure 2.5. Truncation of continuous signal.

2.2.4 Power spectral density (P.S.D.)

If we now return to the comparison with the Fourier series and regard $x_T(t)$ as a repetitive signal of period $2T$, then it would have a **line spectrum.** The power associated with a particular frequency component $X_r\mathrm{e}^{\mathrm{i}r\omega_0 t}/2T$ is proportional to $|X_r|^2/4T^2$ and lies in the frequency range $r\omega_0/2\pi$ to $(r+1)\omega_0/2\pi$. Hence the power per unit frequency interval is

$$\frac{|X_r|^2/4T^2}{[(r+1)\omega_0/2\pi]-[r\omega_0/2\pi]} = \frac{1}{2T}|X_r|^2\,.$$

It is reasonable therefore to define a function which, as we shall see later, is the ratio of power to unit bandwidth, as follows:

$$S_{xx}(\omega) = \lim_{T\to\infty}\frac{1}{2T}|X_T(\mathrm{i}\omega)|^2 = \lim_{T\to\infty}\frac{1}{2T}X_T(\mathrm{i}\omega)X_T^*(\mathrm{i}\omega)\,, \tag{2.10}$$

where the asterisk denotes the complex conjugate. As $T\to\infty$ the line spectrum merges into a continuous curve known as the **power spectral density** (P.S.D.).

Power in any particular band, from ω_1 to ω_2 say, is simply

$$\tfrac{1}{2}\pi\int_{\omega_1}^{\omega_2} S_{xx}(\omega)\mathrm{d}\omega\,.$$

2.2.5 Parseval's theorem

We may put the concept of power spectral density on a firmer footing by noting that Parseval's theorem states that

$$\lim_{T\to\infty} \frac{1}{2T}\int_{-T}^{T} x^2(t)\mathrm{d}t = \frac{1}{2\pi}\int_{-\infty}^{\infty} S_{xx}(\omega)\mathrm{d}\omega \,. \tag{2.11}$$

The left hand side of the above expression is simply the mean square of the signal or total power. The use of the word power comes from its original use in electrical work when $x(t)$ was considered to be a voltage or a current. It is now generally used even when $x(t)$ has a mean square which is not strictly a power in the same sense. The left hand side will have dimensions V^2, I^2, T^2, velocity^2, etc.

The right hand side of the expression in Parseval's theorem is also total power and hence $S_{xx}(\omega)$ is a power density or a measure of how the power is 'spread out' in frequency. Therefore $S_{xx}(\omega)$ is called the **power spectral density** and has units W Hz^{-1}, or alternatively V^2 Hz^{-1}, A^2 Hz^{-1}, K^2 Hz^{-1}, vel^2 Hz^{-1}, m^2 s^{-4} Hz^{-1}, etc.

2.2.6 Block diagram of a P.S.D. analyser

The block diagram of figure 2.6 indicates how a signal $x(t)$ can be either summarised by its mean square value, or decomposed by being fed through a number of parallel filters.

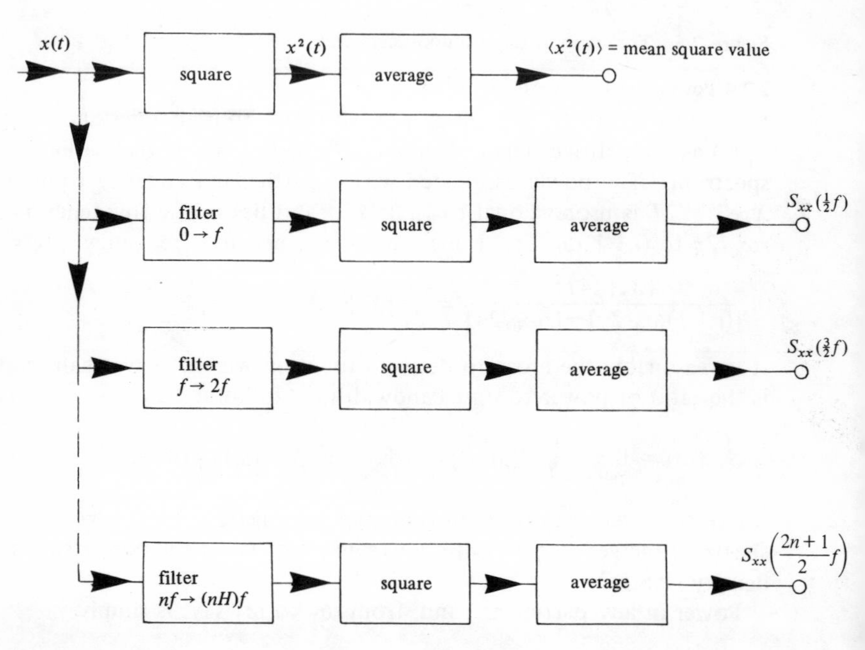

Figure 2.6. Block diagram of a spectrum analyser.

The outputs $S_{xx}\left(\frac{2n+1}{2}f\right)$, where $n = 0, 1, 2 ...$, are simply ordinates on the P.S.D. curve. The discussion of some practical ways of obtaining the P.S.D. is given in chapter 3.

2.3 The level crossing problem

An alternative way of considering the behaviour of a random signal is to examine the average frequency with which that signal crosses over a given amplitude level. The level normally considered is zero and the problem is then called the zero crossing problem. There is, however, no need to restrict the consideration to the zero level, and we therefore consider the probability of the signal crossing a set level a with positive slope.

Now the probability that $x(t)$ will cross the level $x = a$ with positive slope in time δt places restrictions both on $x(t)$ and on its slope $\dot{x}(t)$. This is demonstrated in figure 2.7.

These restrictions are

$$-\infty < x(t) < a \quad \text{and} \quad \frac{a - x(t)}{\delta t} < \dot{x}(t) < \infty .$$

Before we can proceed further it will be necessary to restrict our attention to a stationary process, and show that in a stationary process $x(t)$ and $\dot{x}(t)$ are uncorrelated (the dot indicates a derivative with respect to time).

For a stationary process we have

$$\frac{1}{2T}\int_{-T}^{T} x^2(t)\,dt = \text{const.} \tag{2.12}$$

Differentiating with respect to time we obtain

$$\frac{1}{2T}\int_{-T}^{T} 2x(t)\dot{x}(t)\,dt = 0 , \tag{2.13}$$

i.e. $x(t)$ and $\dot{x}(t)$ are uncorrelated; we shall assume that $x(t)$ has a p.d.f. f(x), and $\dot{x}(t)$ has a p.d.f. p($\dot{x}$).

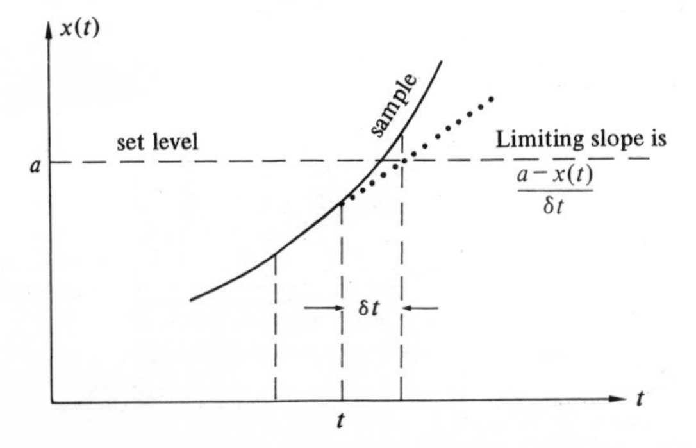

Figure 2.7. Illustrating probability of crossing $x = a$ in time δt.

The probability that $-\infty < x(t) < a$ is

$$\int_{-\infty}^{a} f(x)\,dx\ ,$$

which is the shaded area on the p.d.f. of $x(t)$, as shown in figure 2.8. The probability that $\frac{a-x(t)}{\delta t} < \dot{x}(t) < \infty$ is given by the integral

$$\int_{\frac{a-x(t)}{\delta t}}^{\infty} p(\dot{x})\,d\dot{x}\ ,$$

which is the shaded area on the p.d.f. of $\dot{x}(t)$, as shown in figure 2.8. Since $x(t)$ and $\dot{x}(t)$ are uncorrelated, the probability that $x(t)$ will cross the level a in time δt is simply the product of the two separate probabilities, and the frequency of crossing is obtained by dividing by δt. Thus, the crossing probability is given by

$$\phi_T(a) = \frac{1}{\delta t}\int_{-\infty}^{a}\int_{\frac{a-x(t)}{\delta t}}^{\infty} f(x)p(\dot{x})\,dx\,d\dot{x}\ . \tag{2.14}$$

The field of integration is sketched in figure 2.9. We see that

$$\phi_T(a) = \frac{1}{\delta t}\int_{0}^{\infty}\int_{a-\dot{x}\delta t}^{a} f(x)p(\dot{x})\,dx\,d\dot{x}\ . \tag{2.15}$$

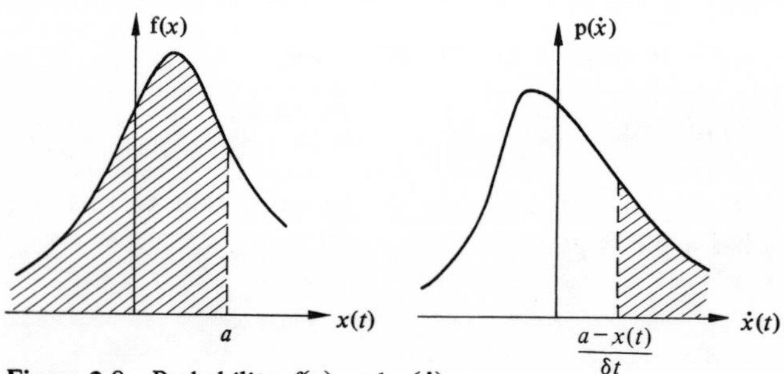

Figure 2.8. Probability $f(x)$ and $p(\dot{x})$.

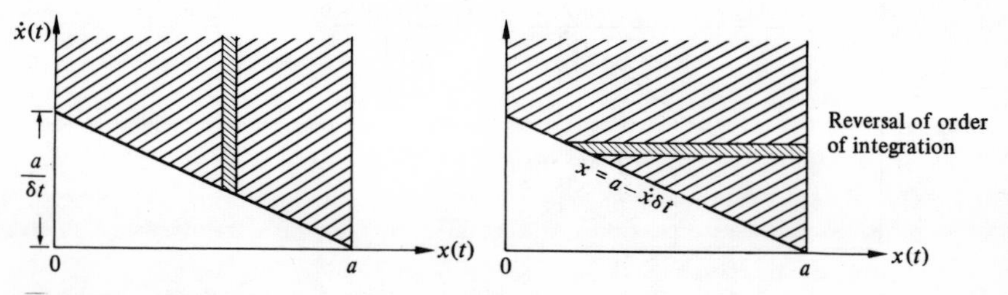

Figure 2.9. Field of integration showing change of variable.

The integral with respect to x is shown graphically in figure 2.10.

The following simple result then follows:

$$\phi_T(0) = \mathrm{f}(a)\int_0^\infty \dot{x}\mathrm{p}(\dot{x})\,\mathrm{d}\dot{x}\ . \tag{2.16}$$

This is a general result and quite independent of the form of $\mathrm{f}(x)$ and $\mathrm{p}(x)$. If, however, these p.d.f's are Gaussian, then further simplification is possible. For example assume

$$\mathrm{f}(x) = \frac{1}{\sigma_x(2\pi)^{1/2}}\exp\left(-\frac{x^2}{2\sigma_x^2}\right)$$

and

$$\mathrm{p}(\dot{x}) = \frac{1}{\sigma_x(2\pi)^{1/2}}\exp\left(-\frac{\dot{x}^2}{2\dot{\sigma}_x^2}\right).$$

Then

$$\phi_T(a) = \frac{1}{\sigma_x(2\pi)^{1/2}}\exp\left(-\frac{a^2}{\sigma_x^2}\right)\frac{1}{\sigma_x(2\pi)^{1/2}}\int_0^\infty \dot{x}\exp\left(-\frac{\dot{x}^2}{2\dot{\sigma}_x^2}\right)\mathrm{d}x$$

$$= \frac{\dot{\sigma}_x}{\sigma_x}\frac{1}{2\pi}\exp\left(-\frac{a^2}{\dot{\sigma}_{\dot{x}}^2}\right). \tag{2.17}$$

The zero crossing frequency is given by

$$\phi_T(0) = \frac{\dot{\sigma}_x}{\sigma_x}\frac{1}{2\pi}\ . \tag{2.18}$$

As a simple example consider $x(t) = A\sin\omega t$, which has a σ_x or r.m.s. value $= A/2^{1/2}$, and $\dot{x}(t) = A\omega\cos\omega t$ with $\dot{\sigma}_{\dot{x}} = A\omega/2^{1/2}$. Then

$$\phi_T(0) = \frac{A\omega}{2^{1/2}}\frac{2^{1/2}}{A}\frac{1}{2\pi} = f \text{ as expected}\ .$$

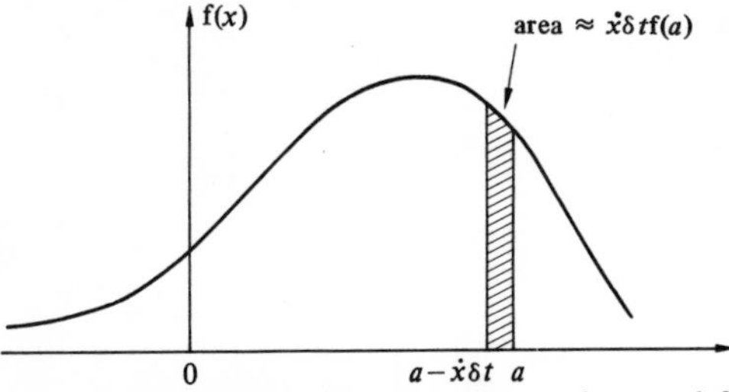

Figure 2.10. Graphical representation of integral for crossing probability.

2.4 Analysis in the time domain

2.4.1 Correlation functions

As we have seen, a simple linear system may be described in the frequency domain by $H(\omega)$ or in the time domain by $h(t)$; $H(\omega)$ and $h(t)$ are Fourier transform pairs.

A signal may also be described in either domain. In the frequency domain the description is by $S_{xx}(\omega)$ and, as one might expect, by its transform in the time domain. The function in the time domain is called the **autocorrelation function** and is defined as

$$R_{xx}(\tau) = \lim_{T\to\infty} \frac{1}{2T}\int_{-T}^{T} x(t)x(t-\tau)\mathrm{d}t \,. \tag{2.19}$$

With this definition it may be shown that

$$S_{xx}(\omega) = \int_{-\infty}^{\infty} R_{xx}(\tau)\mathrm{e}^{-\mathrm{i}\omega\tau}\mathrm{d}\tau \tag{2.20}$$

and

$$R_{xx}(\tau) = \frac{1}{2\pi}\int_{-\infty}^{\infty} S_{xx}(\omega)\mathrm{e}^{\mathrm{i}\omega\tau}\mathrm{d}\omega \,. \tag{2.21}$$

These are known as the Wiener-Khintchine relationships and appear in a variety of forms in the literature. The old fashioned name for $R_{xx}(\tau)$ was the **periodogram**, a very descriptive title as it fully explained one of the main uses of $R_{xx}(\tau)$, namely to recover periodic signals 'buried' in noise.

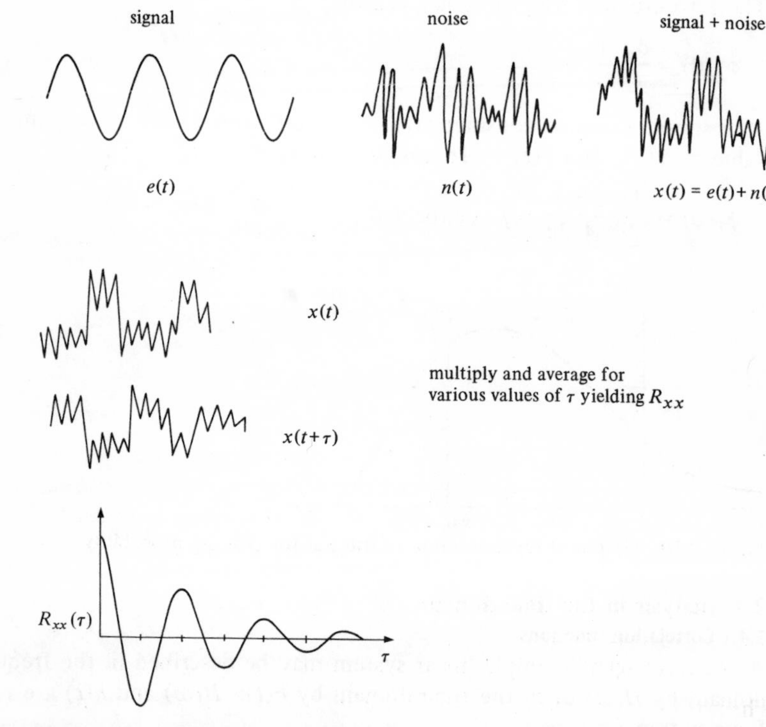

Figure 2.11. Compilation of $R_{xx}(\tau)$.

2.4.2 Graphical demonstration of the compilation of $R_{xx}(\tau)$

The essential steps in the compilation of $R_{xx}(\tau)$ are shown in figure 2.11. Note that $R_{xx}(0)$ is simply the mean square of the signal, $\overline{x^2(t)}$, and this value is sometimes used as a normalising factor to yield the **correlation coefficient**

$$\sigma_{xx}(\tau) = \frac{R_{xx}(\tau)}{R_{xx}(0)} . \tag{2.22}$$

Autocorrelation can recover signals with an S/N ratio about -40 dB.

2.4.3 Block diagram of an $R_{xx}(\tau)$ analyser

Figure 2.12 gives a diagram of a simple analyser to calculate values of the autocorrelation coefficient.

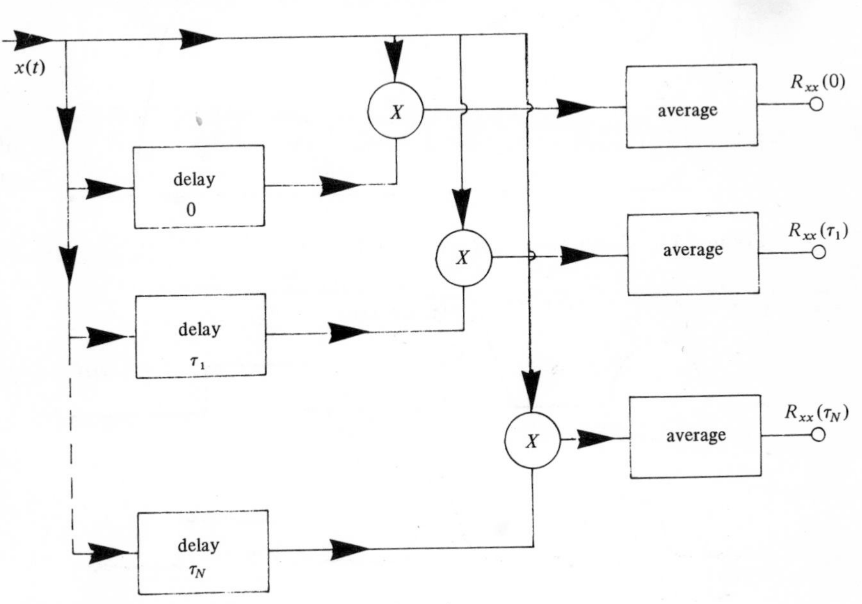

Figure 2.12. Block diagram of an $R_{xx}(\tau)$ analyser.

2.4.4 Graphical comparison of $R_{xx}(\tau)$ and $S_{xx}(\omega)$

In the example given in figure 2.12 the information about the signal $x(t)$ may be displayed in *either* the time domain *or* the frequency domain, as shown in figure 2.13. It is emphasised that the information content is invariant.

Let us define

$$R_{xy}(\tau) = \lim_{T\to\infty} \frac{1}{2T}\int_{-T}^{T} x(t)y(t-\tau)\mathrm{d}t . \tag{2.23}$$

The most obvious application of this function is in finding the delay between two similar signals. Such a pair of signals is illustrated in figure 2.14.

If we delay $y(t)$ by an amount τ and multiply by $x(t)$ and average, repeating for various values of τ, we obtain the graph of figure 2.15.

The peak in the correlogram occurs at a delay T equal to the actual delay in the two signals. The noise on signal $y(t)$ is uncorrelated with $x(t)$ and hence averages to zero.

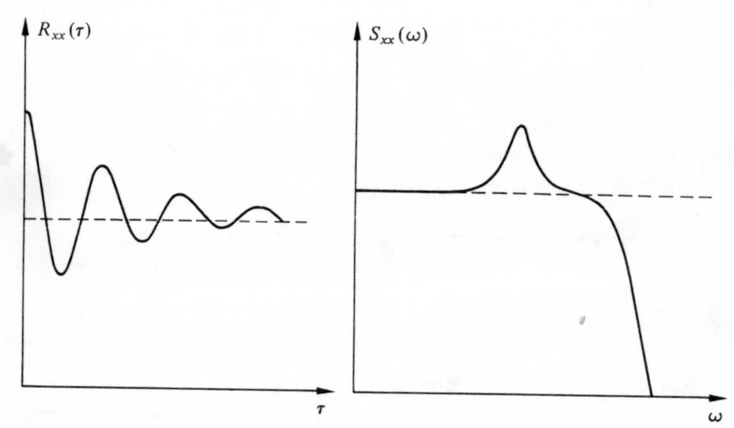

Figure 2.13. Comparison of $R_{xx}(\tau)$ and $S_{xx}(\omega)$.

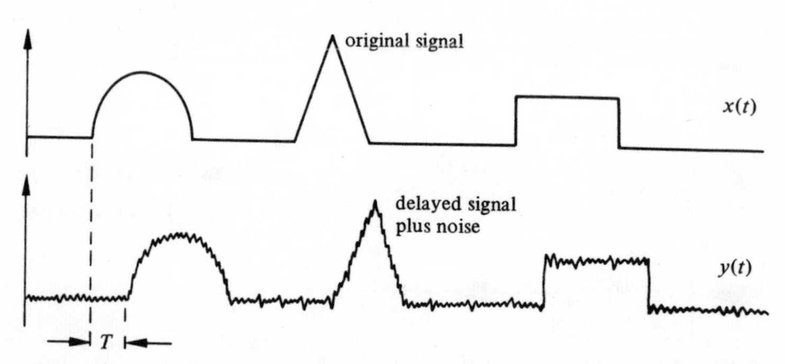

Figure 2.14. Two signals differing by a time delay.

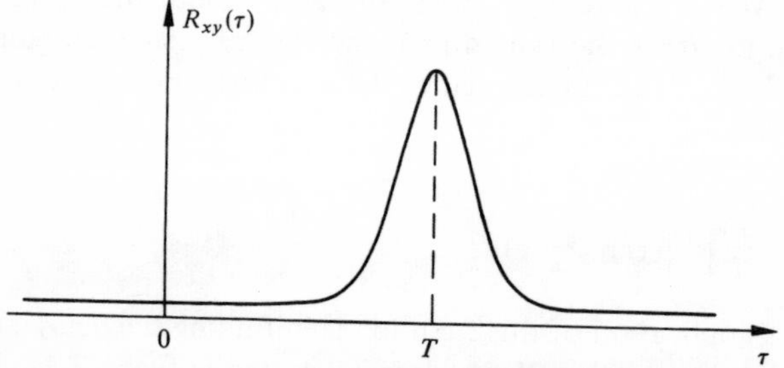

Figure 2.15. Cross correlogram for signals of figure 2.14.

The **degree of correlation** is given by

$$\sigma_{xy}(\tau) = \frac{R_{xx}(\tau)}{R_{xy}(\tau)}, \qquad 0 < |\sigma| < 1 . \tag{2.24}$$

$|\sigma| = 1$ perfect correlation,
$|\sigma| = 0$ no correlation.

$R_{xy}(\tau)$ has also a Fourier transform

$$S_{xy}(\omega) = \int_{-\infty}^{\infty} R_{xy}(\tau) e^{-i\omega\tau} d\tau . \tag{2.25}$$

$S_{xy}(\omega)$ may also be defined in a similar manner to $S_{xx}(\omega)$ as

$$S_{xy}(\omega) = \lim_{T\to\infty} \frac{1}{2T} X_T(i\omega) Y_T^*(i\omega) . \tag{2.26}$$

$S_{xy}(\omega)$ is the **cross spectral density** and measures the relationship between a pair of signals—the frequency domain.

2.4.5 Calculation of cross power spectral density

Figure 2.16 gives a block diagram of an analyser for calculating the cross P.S.D.

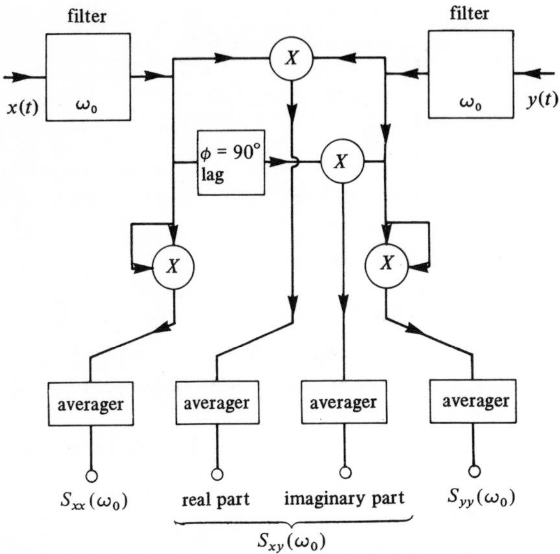

Figure 2.16. Block diagram of cross P.S.D. analyser.

2.5 Random noise

2.5.1 White noise

White noise is a useful mathematical concept, and is defined as a variable having a Gaussian amplitude density distribution and equal power in each unit of frequency increment for all frequencies. White noise therefore has a constant power spectral density from minus infinity to plus infinity. Such a function would have infinite power and clearly cannot exist in practice. However, in some applications a signal may be 'white' over the range of frequencies of interest. The term white noise arises from an analogy with white light, although white light has equal power per unit wavelength and not unit frequency as implied in white noise.

If we consider the case where the signal is white, i.e. the power spectral density has a constant amplitude A, then

$$S_{xx}(\omega) = A \, . \tag{2.27}$$

Applying the Wiener–Khintchine relationships we find that the autocorrelation function is given by

$$R_{xx}(\tau) = \frac{1}{2\pi}\int_{-\infty}^{\infty} A\,\mathrm{e}^{\mathrm{i}\omega\tau}\,\mathrm{d}\omega = A\,\delta(\tau) \, . \tag{2.28}$$

That is, the function is only correlated at the origin and completely uncorrelated elsewhere. A sketch of the power spectral density and the autocorrelation function for white noise is given in figure 2.17.

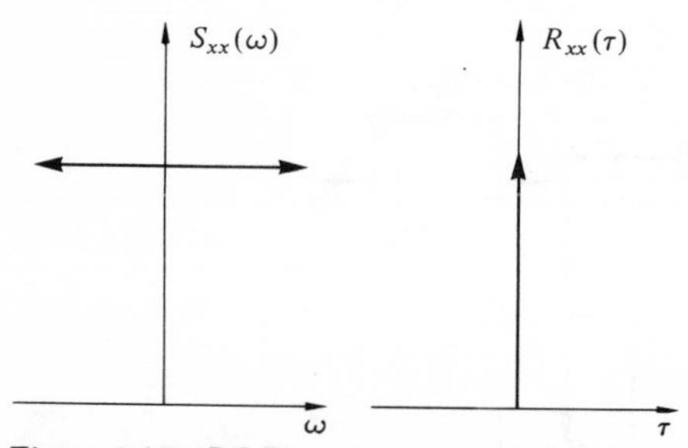

Figure 2.17. P.S.D. and autocorrelation for white noise.

2.5.2 Band-limited white noise

Band-limited white noise is sometimes a convenient representation of a practical situation in which the power is constant between two values of ω and sensibly zero elsewhere.

We consider the case when the power spectral density of a signal has the following properties:

$$S_{xx}(\omega) = A \, , \qquad \omega_1 \leqslant \omega \leqslant \omega_2$$

$$S_{xx}(\omega) = 0 \qquad \text{elsewhere} \, .$$

Then, as before, we may calculate the autocorrelation function by the Wiener–Khintchine relationship:

$$R_{xx}(\tau) = \frac{1}{2\pi}\int_{-\infty}^{\infty} S_{xx}(\omega)\,\mathrm{e}^{\mathrm{i}\omega\tau}\,\mathrm{d}\omega \, . \tag{2.29}$$

As $S_{xx}(\omega)$ is an even function we may write this as

$$R_{xx}(\tau) = \frac{1}{\pi}\int_{\omega_1}^{\omega_2} A\cos\omega\tau\,\mathrm{d}\omega = \frac{A}{\pi\tau}(\sin\omega_2\tau - \sin\omega_1\tau)\,. \tag{2.30}$$

Two particular cases of interest now arise:

(i) $\omega_1 = 0$, in which case the process is the same as if pure white noise was passed through a low pass filter

(ii) $\omega_2 = \omega_1 + \delta\omega$, a very narrow pass band of width $\delta\omega$.

In case (i) the autocorrelation of low pass white noise reduces to

$$R_{xx}(\tau) = \frac{A\omega_2}{\pi}\frac{\sin\omega_2\tau}{\omega_2\tau}\,. \tag{2.31}$$

This function and the P.S.D. from which it arises are shown in figure 2.18. In the case (ii) the autocorrelation of the narrow band white noise becomes

$$R_{xx}(\tau) = \frac{A}{\pi\tau}[\sin(\omega_1 + \delta\omega)\tau - \sin\omega_1\tau]\,.$$

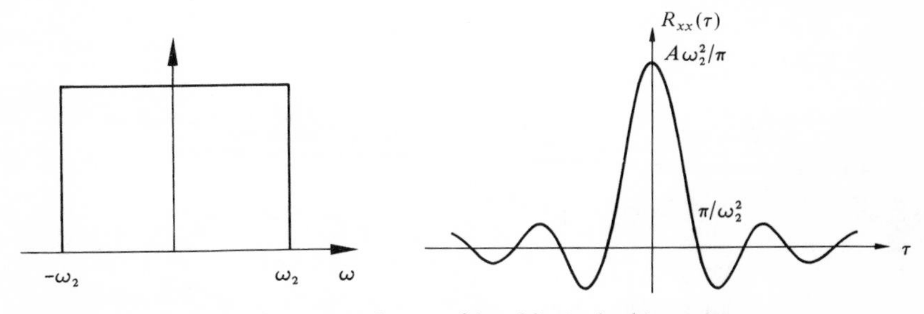

Figure 2.18. P.S.D. and autocorrelation of band-limited white noise.

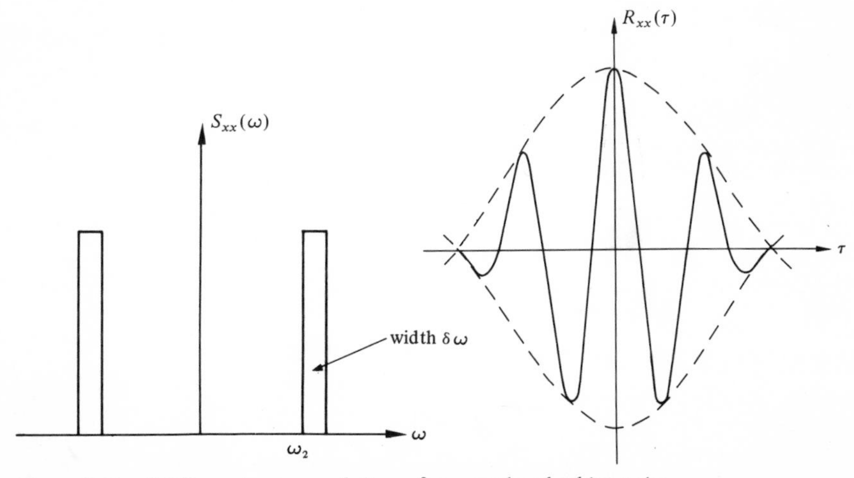

Figure 2.19. P.S.D. and autocorrelation of narrow band white noise.

Assuming $\cos\delta\omega = 1$ we have

$$R_{xx}(\tau) = \frac{A\delta\omega}{\tau}\cos\omega_1\tau\frac{\sin\delta\omega\tau}{\delta\omega\tau} \quad . \tag{2.32}$$

An examination of this expression shows that the autocorrelation is simply a cosine wave of frequency ω_1 which is amplitude modulated by the term $\sin\delta\omega\tau/\delta\omega\tau$. The P.S.D. and correlation function for narrow band white noise are shown in figure 2.19.

3

Measurement of random signals

3.1 Introduction

In the analysis of random signals one can employ either an analogue, hybrid analogue/digital, or pure digital technique, or some combination of these, and the analysis may be carried out in either a time or a frequency domain. We shall first consider the various analogue devices, then the hybrid devices, and finally the pure digital devices.

When considering the frequency analysis of a signal we are always confronted with a form of uncertainty principle which has to be taken into consideration. If we wish to obtain a high accuracy in our determination of the amplitude of a signal at a particular instant, then the response of our analysing device must pass a wide band of frequencies, and so we are unable to define accurately the frequency at which we are making the measurements. Alternatively, if we wish to specify precisely the frequency at which we are making measurements, then our analysing device will require a long time to stabilise, and thus we are unable to say with great precision what the amplitude will be at a particular instant. Thus precision in both amplitude and frequency determinations at a particular instant is not possible, and a compromise must be struck. A partial answer to the dilemma can be achieved by using 'ensemble averaging' which will be discussed more fully in a later section.

One might wish that filters in an analyser would have an effective response as a function of frequency like the transfer function shown in figure 3.1(a); however, such a response is difficult to achieve (certainly impossible using analogue techniques), and has an unsatisfactory performance for transient signals, as a form of 'ringing' occurs. The usual response of the filter is like the transfer function of figure 3.1(b). This type of function can be produced more easily using analogue techniques and responds in a more satisfactory manner to transient

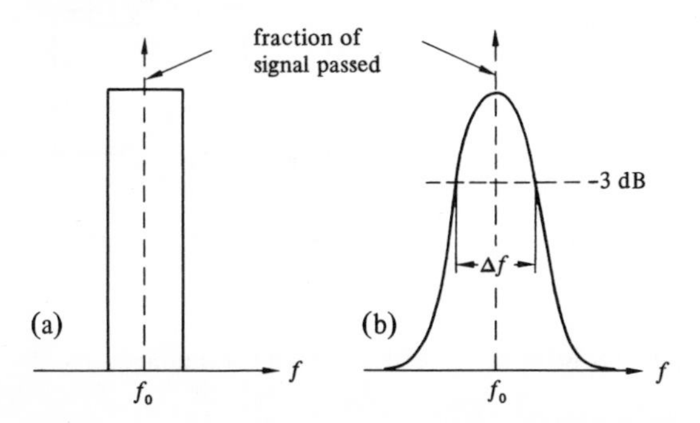

Figure 3.1. Filter characteristics: (a) ideal filter transfer function; (b) real filter transfer function.

signals. It is felt that the shape of the transfer function should be 'Gaussian' [1] for the best transient response. The transfer function in figure 3.1(b) has marked on it the 3 dB points which are employed to indicate the range of frequencies passed by the filter. The 3 dB points are those where the energy passed drops to 3 dB ($=\frac{1}{2}$) of the maximum value, or the amplitude drops to 0·707 ($=\sqrt{\frac{1}{2}}$); since energy $\propto$ (amplitude)^2.

Another problem that faces us is to decide the form of the output from our analysing device. The usual display modes are peak–peak, root-mean-square (r.m.s.) and the logarithm of the previous two (to obtain a dB rating). The presentation of a peak–peak display does not present many difficulties and can be achieved using a simple rectifier–capacitor combination as shown in figure 3.2. Notice that the resistance of the meter determines the time constant of the system, and other external resistors are needed if a different time constant is required. There are problems associated with obtaining r.m.s. and logarithmic outputs using analogue devices, which can be more simply overcome in a digital system. The significance of the r.m.s. output mode is that the average power is proportional to the (r.m.s.)^2. The simplest way to achieve an r.m.s.

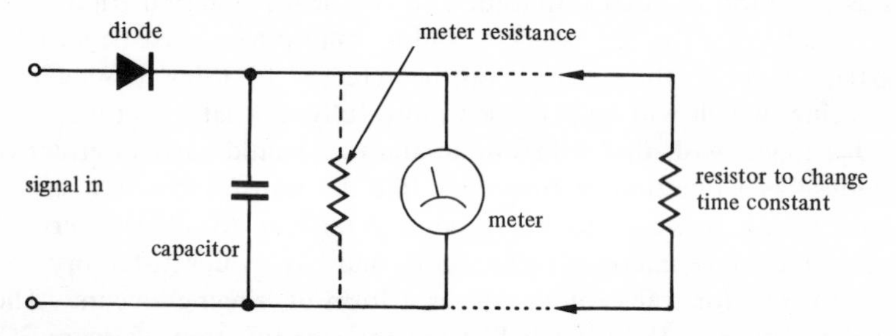

Figure 3.2. Circuit for peak–peak indication.

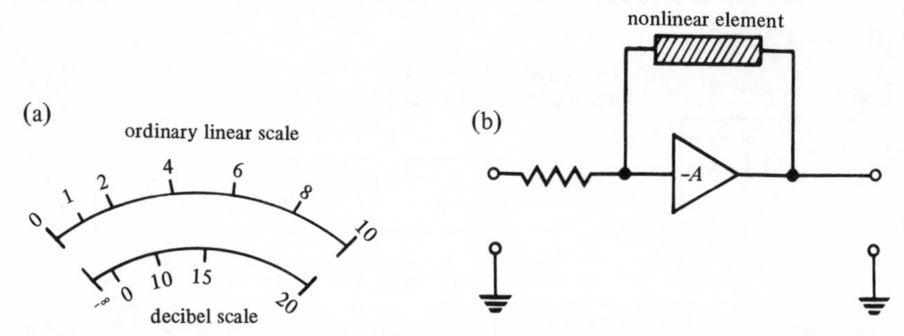

Figure 3.3. Arrangements for logarithmic (dB) indication: (a) nonlinear scale; (b) logarithmic converter.

(1) A Gaussian characteristic is of the form $A\exp[-\beta(f-f_0)^2]$ where f is frequency, f_0 is the centre frequency of the filter, and β is some constant.

indication is to use the fact that for signals at a specific frequency r.m.s. = (peak–peak)/($2\sqrt{2}$); however, for signals containing a range of frequencies this simple relationship no longer applies and hot wire bolometers must then be used. A logarithmic indication can be achieved by using a nonlinear scale on a meter as shown in figure 3.3(a); however, if a voltage proportional to the logarithm of a signal is required, then an operational amplifier employing a nonlinear feedback element having a logarithmic characteristic can be employed as shown in figure 3.3(b). This logarithmic characteristic of the nonlinear element can be synthesised by approximating the characteristic by short linear elements and then using diode switches.

There is also another form of representation of the output that we may have to consider. Often the signals being analysed are impulsive or burst-like and then the quantity of importance is the maximum amplitude that has occurred, and this measurement has to be stored. Thus we require another form of output—an impulse mode—and this can simply be achieved by a modification of the circuit in figure 3.2 as shown in figure 3.4. In the circuit of figure 3.4 the voltage stored in the capacitor will correspond to the maximum amplitude of the signal applied to the diode, since there is a negligible leakage of charge to earth via the high resistance metering circuit (the time constant of the capacitor/metering circuit is extremely long).

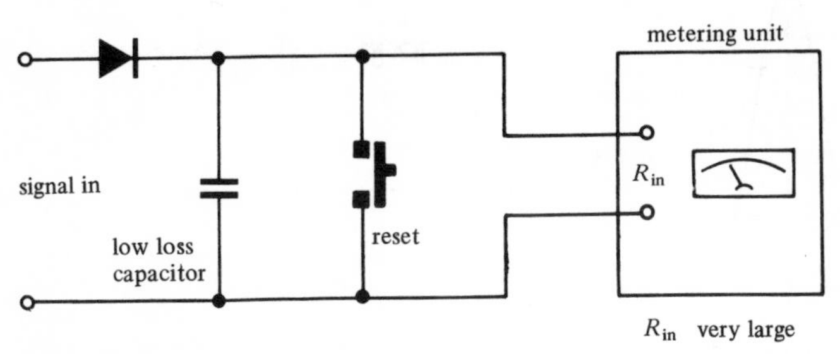

Figure 3.4. Circuit for impulse measurements.

3.2 Analogue methods

3.2.1 Analogue frequency analysers

3.2.1.1 *Introduction*

There are two basic types of analogue frequency analyser; the serial and the parallel analyser. In the serial analyser the contribution at each frequency is measured consecutively, or serially, and this means that the signal must be repeated many times for each frequency in the analysis. The parallel analyser has a bank of filters which are excited simultaneously and whose outputs are all available simultaneously. A parallel analyser has obvious advantages for use in the analysis of transients, where the signal may occur only once and cannot be simply repeated many times for analysis. However, the physical complexity associated with several hundred filters, say, and the provisions for simultaneous read out of all filters, imposes limitations on this approach.

There are two types of filters available for use in frequency analysers: purely electrical and electromechanical. A simple inductance/capacitance (L/C) combination has a filter characteristic similar to that in figure 3.5, and the bandwidth can be controlled by a resistance (R). The centre frequency of such a filter can be altered by changing the value of either the inductance and/or the capacitance. The electromechanical filter uses the fact that a solid can vibrate at only certain frequencies depending on the nature of the solid and its dimensions. To utilise a mechanical element as a filter some means must be employed to convert electrical signals into mechanical vibrations and back again (transducers), but fortunately both the transducer and the mechanical element are combined if one uses the piezoelectric properties of quartz, as shown in figure 3.6. Such an electromechanical filter is a fixed-frequency device (the centre frequency of the filter cannot be altered).

In certain applications a filter characteristic having a flat top and passing a wider range of frequencies is employed. This is known as a band-pass filter and will have a characteristic like that of figure 3.7(a). This type of

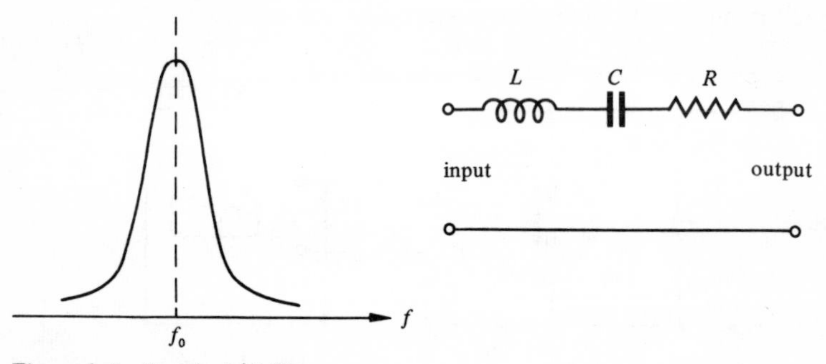

Figure 3.5. Simple L/C filter.

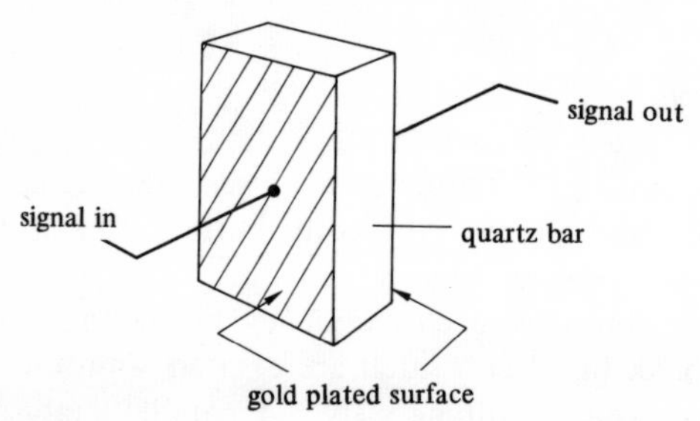

Figure 3.6. Simple electromechanical filter.

filter is constructed by using complex configurations of inductive and capacitive elements; for example a band-pass filter may be made from a combination of low-pass and high-pass filters whose characteristics are shown in figure 3.7(b) and (c) together with typical circuits.

3.2.1.2 *Serial analogue frequency analysers*
The simplest form of serial analyser would employ an *LC* tuned circuit (figure 3.5) and the centre frequency of the filter would be varied by varying the capacitance. Unfortunately the range of variation of frequency is limited to about 2 to 1 due to limitations on the possible variation of capacitance. Thus, to analyse a wide spread of frequencies many range changes are required (switching-in a different inductance). Also, the bandwidth of the circuit varies as the centre frequency is changed, becoming narrower with increasing centre frequency. It is customary to have either a fixed bandwidth or fixed percentage bandwidth (bandwidth/centre frequency = constant), and the simple *LC* circuit does not achieve either of these two requirements. The use of more sophisticated electrical networks can provide a fixed percentage bandwidth analyser, a typical bandwidth being 5%.

Probably the most useful type of frequency (or wave, or spectrum) analyser is one where the bandwidth is constant and narrow, say of the order of several hertz, and also where the frequency setting can be varied considerably without any range changes. These requirements are best met

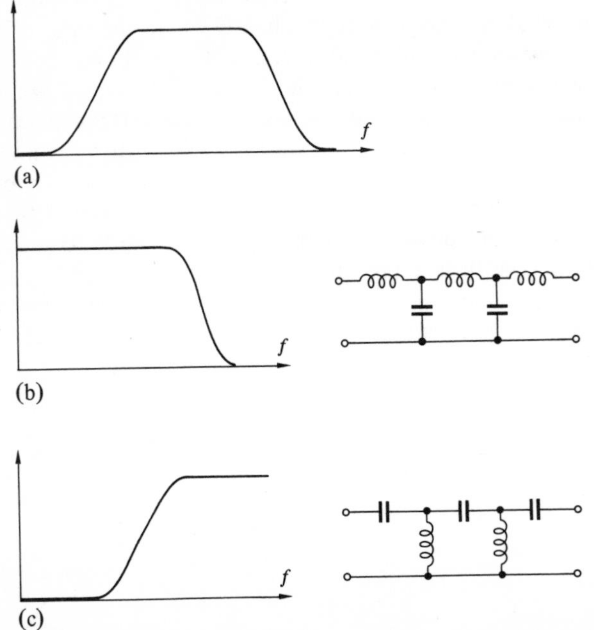

Figure 3.7. Filters: (a) band-pass; (b) low-pass; and (c) high-pass.

by a heterodyne-type analyser. A block diagram of such an analyser is given in figure 3.8 and the route of a 1 kHz signal to which the analyser has been set is shown. The signal is first mixed with a 49 kHz signal from an oscillator so that outputs at 49 kHz and 49 ± 1 = 50, 48 kHz are produced. Only the signal at 50 kHz is allowed through the narrow bandwidth crystal filter. The 50 kHz signal from the filter is again mixed with the 49 kHz signal from the oscillator (this step is not always carried out) so that frequencies of 49 ± 50 = 1, 99 kHz are produced at the output of the mixer. The signal is now passed through a low-pass filter so that only the 1 kHz is seen at the final output. Thus we have isolated any contributions at 1 kHz. If we change the oscillator frequency to say 48 kHz, then we would filter out the contributions at 2 kHz. Notice that instead of using 49 kHz for the oscillator signal we could have used 51 kHz. The heterodyning principle may in fact be used twice with two different crystal filters and two different oscillators so as to achieve superior selectivity.

3.2.1.3 *Parallel analogue frequency analyser*
As mentioned previously there exists both an economic and physical limitation to the number of filters that will be employed in a parallel arrangement. Consequently parallel analysers generally have large bandwidth filter characteristics so that a wide range of frequencies may be analysed. The most important application of parallel filters is the use of a bank of $\frac{1}{3}$-octave filters for acoustical measurements. Audible sound may be considered in what is known as octave bands where the ratio of the frequencies at the centres of the band is 2. Table 3.1 shows the internationally accepted octave frequencies up to 4000 Hz. If we now subdivide each octave into three then we have $\frac{1}{3}$-octave frequencies as also shown in table 3.1. Filters centred on the $\frac{1}{3}$-octave frequencies given in table 3.1 are known as $\frac{1}{3}$-octave filters. These filters are constructed so as to have a constant percentage bandwidth and the exact form of their characteristic has been set down by International Standards Associations. The percentage bandwidth is approximately 20%.

The importance of such $\frac{1}{3}$-octave filters relates to the behaviour of the human ear for it turns out that these filter characteristics delineate the

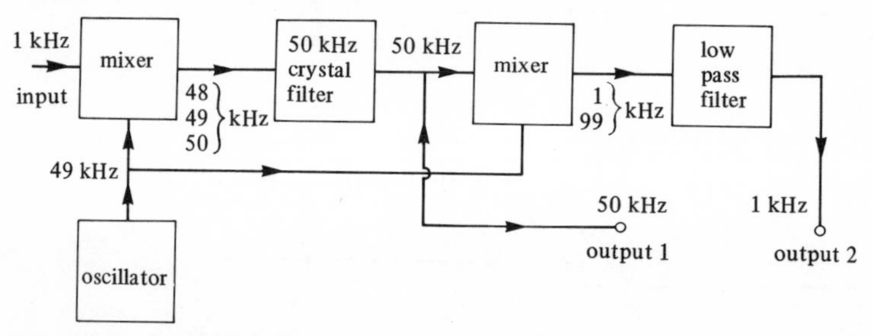

Figure 3.8. Block diagram of a heterodyne frequency analyser.

bands of frequencies which are heard as distinctly different sounds. Any sounds occurring within the range of a $\frac{1}{3}$-octave filter tend to blend together in such a way as to mask each other. The outputs from the parallel $\frac{1}{3}$-octave analyser are often combined in various ways to provide a meaningful indicator of the perceived loudness of a sound.

Table 3.1. Octave and $\frac{1}{3}$-octave frequencies.

Octave frequencies	$\frac{1}{3}$-octave frequencies	Octave frequencies	$\frac{1}{3}$-octave frequencies
31·5	31·5	500	500
	40		630
	50		800
63	63	1000	1000
	80		1250
	100		1600
125	125	2000	2000
	160		2500
	200		3150
250	250	4000	4000
	315		
	400		

3.2.2 Analogue correlators

The physical processes involved in a correlation determination are a variable delay, a multiplication, and a summation. The most difficult process to achieve is the variable delay, and the usual analogue technique

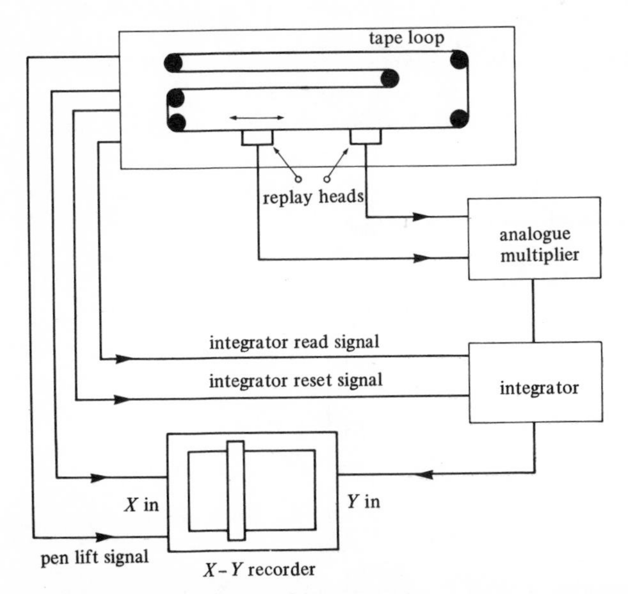

Figure 3.9. Analogue correlator.

employs a special tape recorder where there are two replay heads and the position of one of these can be moved with respect to the other thus providing a variable delay. A schematic diagram of a system for correlation determinations is shown in figure 3.9. The same signal is monitored by both replay heads for autocorrelations and there are two separate signals for cross-correlations. After the entire tape loop has passed the heads the output of the integrator is recorded (usually on an $X-Y$ recorder), the movable head is advanced to a new position, and the integrator is reset to zero in preparation for the determination of the next point of the correlation.

An important function derived from the cross-correlation is the cross-spectral density which is obtained by taking a Fourier transform of the cross-correlation. An alternative way of obtaining the cosine part of the cross-spectral density (co-spectrum) and the sine part of the cross-spectral density (quad-spectrum), is to use two filters, two multipliers, and a 90° phase shifter as shown in figure 3.10.

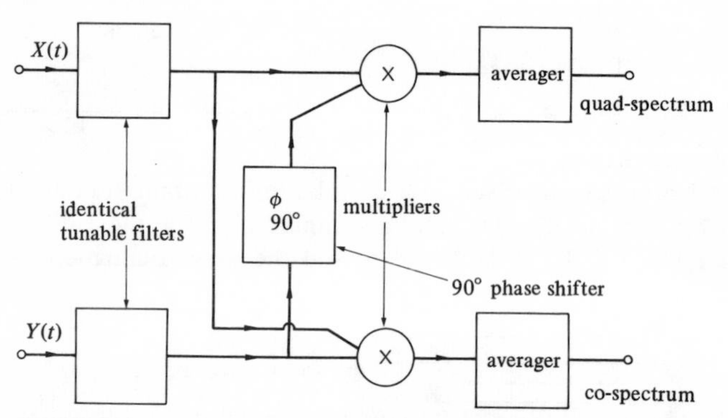

Figure 3.10. Cross-spectral density determination using two identical filters.

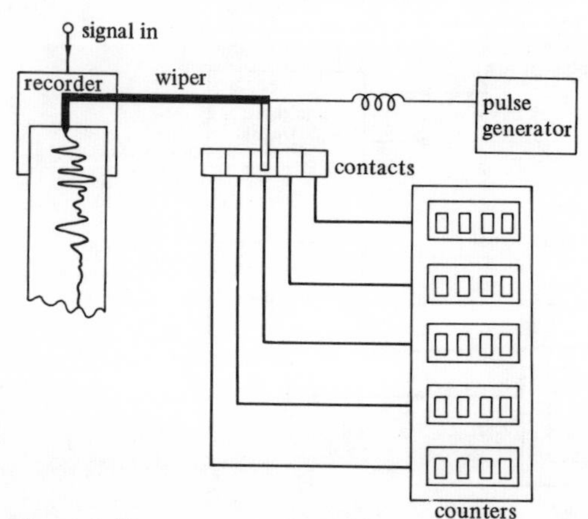

Figure 3.11. Schematic diagram of statistical distribution analyser.

3.2.3 Analogue probability distribution function (p.d.f.) analyser

The determination of the p.d.f. requires some means of determining the amplitude of the signal at a particular instant of time. A simple technique which has been employed is to use a wiper attached to a recorder which slides over a series of contacts. The signal of interest is being displayed on the recorder. A pulse generator producing a regular sequence of pulses is applied to the wiper and thus, depending on which contact the wiper is resting, an appropriate counter corresponding to a given amplitude is incremented. This arrangement is often referred to as a statistical distribution analyser and is shown schematically in figure 3.11. The sorting of the amplitudes and the counting may be done electronically and this greatly decreases the response time of the apparatus.

3.3 Hybrid analogue-digital techniques

3.3.1 Hybrid spectrum analysers

The use of both digital and analogue techniques within one instrument has allowed a great improvement in frequency analysers. We have already seen how a form of uncertainty principle acts during a frequency analysis. Let us imagine that we have signals in the range 0 to 20000 Hz and we wish to analyse these using a filter having a bandwidth of 10 Hz. Now the small 10 Hz bandwidth of the filter means that we cannot pinpoint the spectrum in time to better than $\sim 0 \cdot 1$ s. If the filter bandwidth were 1000 Hz then we could pinpoint the spectrum in time within ~ 1 μs but we would have lost our frequency resolution. However, we could use our 1000 Hz bandwidth filter if the original signal has been speeded up in time by a factor of 100 (or the duration of the signal has been compressed to $0 \cdot 01$ of its original duration) so that the frequency range

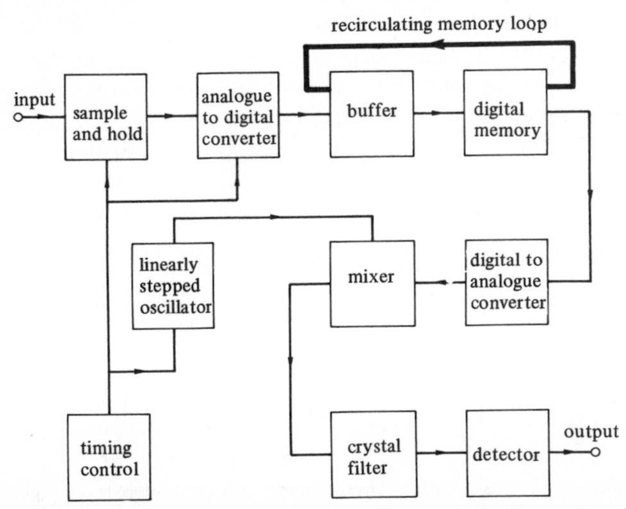

Figure 3.12. Schematic diagram of a frequency analyser using the speed-up technique.

now covers 0 to 2000000 Hz. Thus by speeding up the signal we can get closer to an instantaneous frequency analysis and still end up with narrow effective bandwidths in our filters. The amount of speed up we can achieve by analogue means, such as a variable speed tape recorder, is limited if we want to analyse a reasonable range of frequencies in the original signal. The most effective speed-up technique employs a combination of analogue and digital techniques as illustrated by the basic system of figure 3.12. The signal is fed equidistantly in time by the sample and hold amplifier [see figure 3.13(a)]. These analogue samples are converted to digital samples by the analogue to digital (A/D) converter (see Appendix I for an introduction to binary arithmetic) and then stored in sequence in a digital memory after passing through a buffer unit. Certain precautions have to be observed when sampling a signal by digital means and these will be discussed in detail in section 4. The digitised samples pass sequentially through the memory and then to a recirculating loop. Now, the time interval between successive samples of the input is extremely long compared to that for successive samples for the memory; thus the effect of feeding back samples via the recirculating loop is to compress the signal so that the output from the digital to analogue (D/A) converter is a speeded up signal illustrated in figure 3.13(c).

The speeded up output is now analysed using a heterodyne principle as described in section 2.1.2. However, in this case, the local oscillator is stepped frequency-wise in synchronism with the output from the memory by a master timing control. Notice that this technique allows a transient signal to be stored in the memory and subsequently analysed.

A considerable improvement in the resolution of a frequency spectrum in the presence of noise can be obtained by using the technique of 'ensemble averaging'. If we take many measurements of a spectrum and then average these measurements, we find that the resolution of the spectrum against the background noise is improved. This ensemble averaging is carried out by converting the outputs for the various discrete frequencies from the analyser into digital form and then accumulating these in a digital memory. The various storage locations in the memory

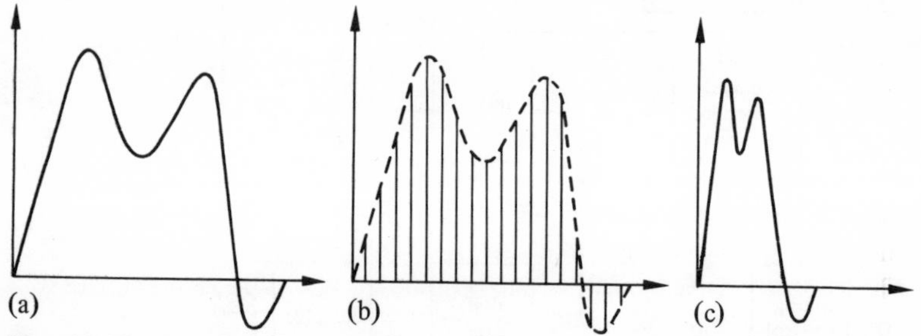

Figure 3.13. Signals at various points in the analyser: (a) input signal; (b) signal to A/D converter; (c) output of D/A converter.

are incremented by the results of each subsequent analysis, so that the final number in each storage location is the sum of the results for all the analyses. Each location of the memory is then read out to give the average spectrum.

3.3.2 Hybrid correlators

The processes required in a correlation determination are the multiplication followed by integration of a signal and a delayed signal (the delay being variable), as shown in figure 3.14. The variable time delay can be realized by digital means, where the signal to be delayed is sampled and digitised, stored and reconstituted from the memory at successively later intervals. The multiplication is then carried out using a conventional analogue multiplier, and the integration using a capacitive integrator.

Actually, having digitised the signal, it is better to perform the operations of multiplication and integration also digitally. Surprisingly, the digitisation of one of the inputs can be quite coarse without impairing the statistical reliability of the results, and one commercial unit uses a seven-bit converter for A input (see figure 3.14) and a three-bit converter for B input (those unfamiliar with the term 'bit' should read Appendix II).

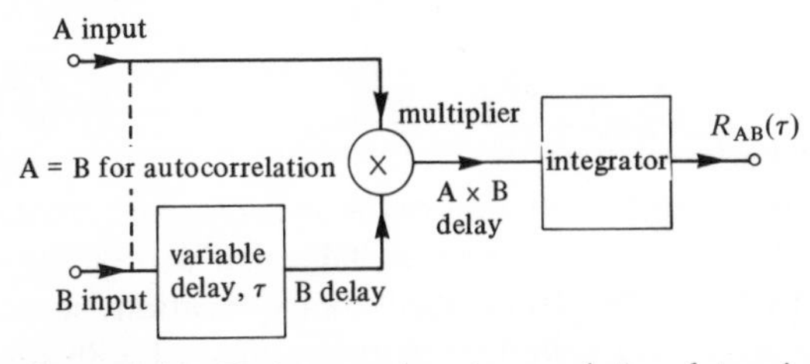

Figure 3.14. Basic operations in correlation determinations.

3.3.3 Other hybrid devices

It is not intended to go into great detail as to the forms of other analysers using combinations of analogue and digital techniques, as the important hybrid devices have been described. Digital techniques have been used to provide storage of results, more precise detector characteristics, and for the production of noise having certain predetermined characteristics for test purposes.

3.4 Digital methods

3.4.1 Introduction

The purely digital approach to signal analysis uses samples of the input signal which are converted to a digital form using an A/D converter, and then all the analyses are performed using digital computer techniques. The computer used may be an ordinary digital computer or a dedicated computer specially designed for signal analysis. The computations are carried out on the data using the available software in a general purpose computer (e.g., Assembler, Fortran, PL1) or by prewired connections in a dedicated computer. There are however certain precautions to be observed in representing data by discrete digitised samples. Figure 3.15 shows that in sampling a signal we are actually replacing it by a sequence of square waves, or we might consider that the signal is being sampled by a rectangular window which moves by discrete steps in time. The consequences and precautions to be observed as a result of this discrete sampling will be considered in section 4.3.

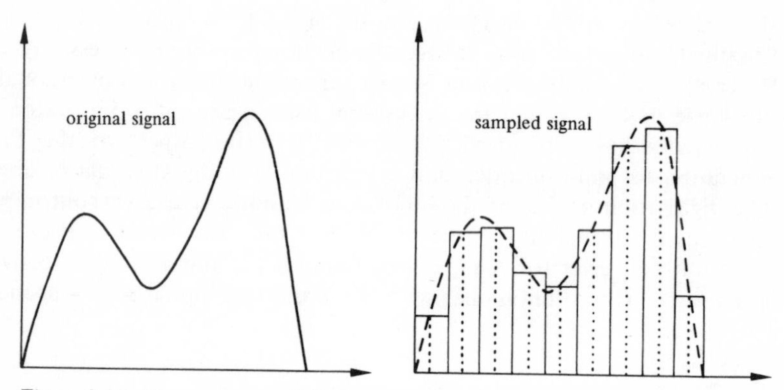

Figure 3.15. Consequence of discrete sampling.

3.4.2 Correlation by digital means

The computation of either autocorrelation or cross-correlation by digital means involves a multiplication of a stored digital signal with another stored digital signal which is taken from later storage locations so as to produce a time delay or lag, and summing of the results. To illustrate, if one signal is stored as amplitudes $A(i)$ $(i = 1, 2, ..., N)$, and another signal as $B(i)$ [for autocorrelation $A(i) = B(i)$], then the lagged products for the correlation are

$$\text{Product at lag } k = \frac{1}{N-k} \sum_{i=1}^{N-k} A(i)B(i+k) . \tag{3.1}$$

The computations involved in equation (3.1) can become rather lengthy as the calculation for each time lag k requires a new retrieval of data from the memory and subsequent operations (multiplication and summing). There is now a tendency in autocorrelation calculations to use the fact that the autocorrelation is the Fourier transform of the power spectral density (P.S.D.), so that the calculation can be speeded up by calculating the P.S.D. using a technique known as the fast Fourier transform (F.F.T.), and then transforming to the autocorrelation again using the F.F.T.

3.4.3 Frequency analysis by digital means

A discrete time series can be considered in the frequency domain by using the discrete Fourier transform (D.F.T.) defined by the equation below

$$\text{D.F.T.} = \frac{1}{N} \sum_{i=1}^{N} A(i) \begin{matrix} \cos \\ \sin \end{matrix} (2\pi \times i \times j/N), \quad j = 1, 2, \ldots, N. \tag{3.2}$$

A considerable number of operations are required to calculate the D.F.T. for all values of j (corresponding to different frequencies)—in fact of the order of N^2 operations—but by the use of the F.F.T. the number of operations is reduced. If the number of data points is a power of 2, then the number of operations required for the F.F.T. is $N \log_2 N$; so for example if $N = 1024 = 2^{10}$, then the D.F.T. requires approximately 10^6 operations while the F.F.T. requires approximately 10^4 operations. An introduction to the F.F.T. is given in Appendix III.

In the consideration of frequency analysis by digital means we become very much aware of the limiting factors imposed by the fact that we are dealing with a sampled waveform. The limitations to be described below will also apply to any type of analysis of sampled data.

We must be on our guard for 'aliasing'. The input signal is sampled at a certain rate, which is often called the strobe frequency, and hence the possibility arises that some frequencies in the input signal may beat with the strobe frequency to produce spurious frequencies within the range being analysed. This is the effect known as aliasing and can be prevented by ensuring that the input signal to the A/D converter is first passed through a low-pass filter designed to cut out all the frequencies which could produce this undesirable effect. An everyday occurrence of aliasing is evident in the phenomenon seen in motion pictures (strobe frequency of 24 Hz as there are 24 frames a second) where such things as spokes in wagon wheels appear to be rotating in the wrong direction.

Another consequence of sampling is that there exists a limit on the upper frequency that may be represented in a frequency analysis. If the time between successive samples is τ, then it is not possible to say anything conclusive about frequencies greater than $1/4\tau$. Figure 3.16 shows a sampling interval of time τ and how frequencies of $1/4\tau$ and $1/2\tau$ are situated with respect to this time interval. It is apparent that

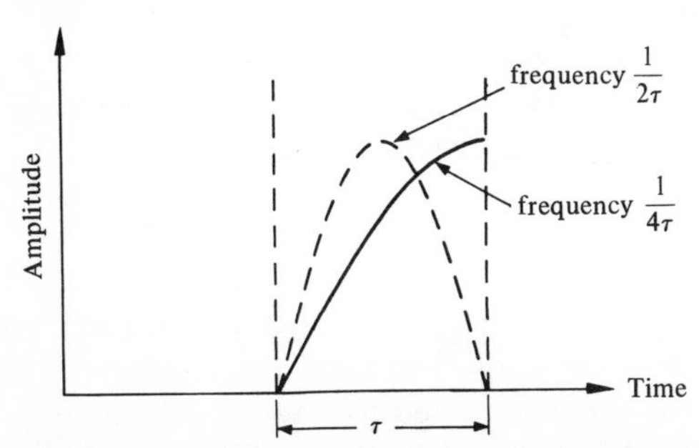

Figure 3.16. Effect of sampling on high-frequency limit.

meaningful information can only be gathered about frequencies less than $1/4\tau$.

A final limitation due to sampling is referred to as frequency 'leakage' and corrections are applied using 'window carpentry'. If we refer back to figure 3.13, then the effect on the frequency spectrum of approximating a continuous waveform by a sequence of square pulses is to cause some distortion, which can be considered due to frequency leakage associated with the Fourier transform of the square pulses. To compensate for this source of error, the data can be modified by viewing them through some mathematical window (this is sometimes referred to as 'prewhitening' the data) or combining the Fourier coefficients obtained from the unmodified data to obtain a new series corrected for sampling errors. There are many forms of window; however, the relative efficacy of the various windows is not really known, so the simplest form of window is usually employed. The simplest window applies a process known as 'hanning' to the Fourier coefficients based on unmodified data to produce a new series. If the Fourier coefficients are $A(i)$ say, then the jth coefficient in the hanned series is

$$A^{\mathrm{H}}(j) = 0{\cdot}25A(j-1)+0{\cdot}5A(j)+0{\cdot}25A(j+1)\,. \tag{3.3}$$

3.4.4 Digital filtering and the z-transform

In digital filtering the digitised samples of a signal are taken and operated on so that the resulting output represents a digitised form of the original signal passed through some filter. The realization of a predetermined filter characteristic using purely analogue means can present some difficulty; however, in theory, the versatility of digital computing techniques is such that they should be able to provide any type of filter characteristic. One obvious technique is to transform the signal into the frequency domain using an F.F.T., multiply the transform by the required frequency response, and then perform an inverse transformation to obtain the filtered signal. This is a roundabout type of process and involves a reasonable amount of computation, even though an F.F.T. has been used to perform the transforms.

Another approach to digital filtering is to combine the digital signals with various weighting factors so that the output is the original signal passed through a filter. This approach to digital filtering arose through a study of the z-transform. If we have a discrete time signal represented by a sequence of numbers x_n, then the z-transform is defined as:

$$X(z) = Z\{x_n\} = \sum_{n=0}^{\infty} x_n z^{-n}\,. \tag{3.4}$$

[Those interested in this approach are referred to the September 1968 issue of the IEEE Transactions on Audio and Electroacoustics.]

A consideration of the z-transform has led to ways of determining coefficients to be used in digital filters. Usually we distinguish between two types of filters; the recursive filter where the current output is computed from the current input and a linear combination of past inputs and outputs, and the non-recursive filter where the current output is computed from the current input and a linear combination of past inputs only. For example, if x_i are inputs, y_i are outputs, and a_i, b_i are filter weights, we have for a recursive filter

$$y_n = a_0 x_n + \dots + a_N x_{n-N} + b_1 y_{n-1} + \dots + b_N y_{n-N}\,, \tag{3.5}$$

and for a non-recursive filter

$$y_n = a_0 x_n + \dots + a_N x_{n-N}\,. \tag{3.6}$$

The emphasis recently has been on the realization of filters using the F.F.T. approach described earlier. However, the realization of filters using a z-transform approach and expressions similar to equations (3.5) and (3.6) may be meaningful when dealing with nonlinear systems and the production of nonlinear filters.

Linear system identification

4.1 Convolution or faltung integral

One of the most interesting problems confronting an investigator is that of identifying the structure of a system when access is limited to applying test signals at the input to the system and observing the resultant at the output. Our attention here will be limited to linear systems which are characterised by either the impulse response or the frequency response. The impulse response $g(\tau)$ and frequency response $G(\omega)$ are formally related as a pair of Fourier transforms as follows:

$$G(\omega) = \int_{-\infty}^{\infty} g(\tau)\mathrm{e}^{-\mathrm{i}\omega\tau}\,\mathrm{d}\tau \tag{4.1}$$

$$g(\tau) = \frac{1}{2\pi}\int_{-\infty}^{\infty} G(\omega)\mathrm{e}^{\mathrm{i}\omega\tau}\,\mathrm{d}\omega\,. \tag{4.2}$$

The impulse response is convenient from an analytical point of view, as any arbitrary function may be regarded as composed of a series of adjacent impulses as shown in figure 4.1.

We see from figure 4.1 that

$$x(t) = x(0)\delta(t)+x(t_1)\delta(t-t_1)+\ldots+x(t_n)\delta(t-t_n)+\ldots\,.$$

The response of a linear system subjected to an arbitrary input may be analysed by considering the input composed of a number of impulses in this way.

The response to the 1st shaded impulse as shown is simply $g(\tau)x(t-\tau)$, and the total output $y(t)$ is the sum of all such impulses:

$$y(t) = \int_{-\infty}^{\infty} g(\tau)x(t-\tau)\,\mathrm{d}\tau\,. \tag{4.3}$$

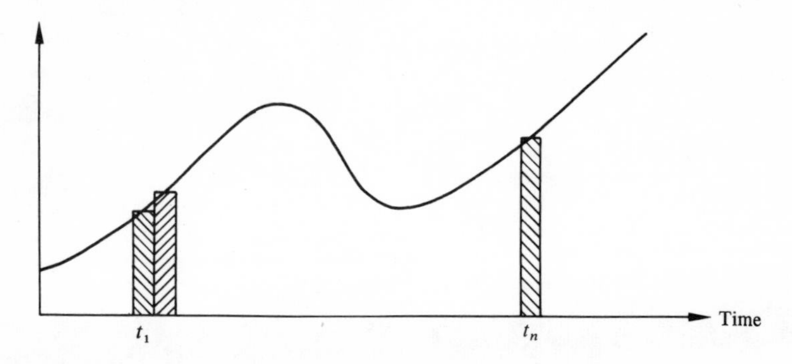

Figure 4.1. Representation of signal by series of adjacent impulses.

For a physically realisable system $g(\tau) = 0$ for $\tau < 0$, and hence for practical systems the lower limit of integration in equation (4.3) is put equal to zero.

The above integral is known as the **convolution** or **faltung** integral and implies that the impulse response function, sometimes called the weighting function, is convolved or folded back along the input signal.

We have seen that, in order to transform the information about the impulse response into the frequency domain, it is necessary to use a Fourier transform. As we shall see, the same transform is useful in system identification problems in transferring from correlation to spectral measurements. We shall however need the result of the transform of a convolution and this is now derived.

4.1.1 Fourier transform of a convolution

Consider

$$y(t) = \int_{-\infty}^{\infty} g(\tau)x(t-\tau)\mathrm{d}\tau \,. \tag{4.4}$$

Taking Fourier transforms of each side we have

$$\int_{-\infty}^{\infty} y(t)\mathrm{e}^{-\mathrm{i}\omega t}\mathrm{d}t = \int_{-\infty}^{\infty}\int_{-\infty}^{\infty} g(\tau)x(t-\tau)\mathrm{e}^{-\mathrm{i}\omega t}\mathrm{d}\tau\,\mathrm{d}t \,.$$

Multiply the right hand side by $\mathrm{e}^{\mathrm{i}\omega\tau}\mathrm{e}^{-\mathrm{i}\omega\tau}$ whose product is unity and hence leaves the integral unchanged. A rearrangement yields

$$\int_{-\infty}^{\infty} y(t)\mathrm{e}^{-\mathrm{i}\omega t}\mathrm{d}t = \int_{-\infty}^{\infty}\int_{-\infty}^{\infty} g(\tau)\mathrm{e}^{-\mathrm{i}\omega\tau}x(t-\tau)\mathrm{e}^{-\mathrm{i}\omega(t-\tau)}\mathrm{d}t\,\mathrm{d}\tau \,.$$

Writing $(t-\tau)$ as a dummy variable in the integral transform, and assuming the order of integration is interchangeable, we see that the right hand side is the product of two Fourier transforms. Hence

$$Y(\omega) = G(\omega)X(\omega) \,. \tag{4.5}$$

In general the Fourier transform of a convolution is equal to the product of the separate transforms.

As a shorthand notation, if we let lower case letters denote the time value of the functions and capital letters Fourier transforms, we can write:

$$a(t) = \int_{-\infty}^{\infty} b(\lambda)c(t-\lambda)\mathrm{d}\lambda$$

or more simply

$$a = b * c \,; \tag{4.6}$$

then

$$A = BC \,.$$

Similarly, if

$$a = (b * c) * d ,$$

then

$$A = BCD .$$

This concept is clearly capable of extension without limit.

4.1.2 Convolution and system identification

In any practical system operating at nominally steady conditions small fluctuations will exist in various operating parameters. The inter-relation between these fluctuations contains information about the dynamics of the system. For example a nuclear reactor constitutes a system in which flow, temperature, and fuel movement perturbations are all related to the reactor power and hence to the state of the coolant. An attractive objective therefore is to examine the possible use of the information contained in these fluctuating signals to monitor the dynamic behaviour of the system.

A further example is the study of the hydrodynamic stability of a boiling channel in which flow, inlet subcooling, and heat flux can be considered as input data. The assumption that such a system is linear is substantially valid for high system pressures. Currently available computer programs for estimating the hydrodynamic stability of boiling coolant systems assume that the system is reducible to a single-input single-output system, and are able to compute the system transfer function. If the reduction to a single-input single-output system is unjustifiable then the system can only be characterised by a transfer matrix in either the time or the frequency domain. The time and frequency concepts are formally equivalent and form Fourier transform pairs. No new information about the system is yielded by transferring from the time to the frequency domain or vice versa. The choice between the two depends on the ease of interpretation and the availability of analysing equipment.

The theory of noise analysis is well established for single-input single-output (SISO) systems. The object of this section is to indicate the possible extension of the analysis to the multiple-input multiple-output (MIMO) systems and indicate the potential of such analysis. For simplicity of presentation the effect of extraneous noise in the measurement is omitted.

4.2 Single-input single-output (SISO) system

A SISO system is shown in figure 4.2. We assume that the system is characterised by either its impulse (delta function) response $g(\tau)$, or its frequency response $G(\omega)$ which form a Fourier transform pair, equations (4.1) and (4.2). The output $y(t)$ is related to the input $x(t)$

by the convolution integral (4.3) which is usually written

$$y = x * g \, .$$

The corresponding expression in the frequency domain is obtained by a Fourier transformation and is

$$Y(\omega) = X(\omega)G(\omega) \, . \tag{4.7}$$

Suppose that the input $x(t)$ is a random fluctuation having a power spectral density defined as before:

$$S_{XX}(\omega) = \lim_{T \to \infty} \frac{1}{2T} X_T(\omega) X_T^*(\omega) \, , \tag{4.8}$$

where the superscript * denotes the complex conjugate, and the suffix T the length of time over which the signal is measured. Similarly:

$$S_{YY}(\omega) = \lim_{T \to \infty} \frac{1}{2T} Y_T(\omega) Y_T^*(\omega) \, . \tag{4.9}$$

Hence it follows, after multiplying each side of equation (4.7) by its complex conjugate, that

$$S_{YY}(\omega) = S_{XX}(\omega) |G(\omega)|^2 \, . \tag{4.10}$$

If the cross-spectral density is further defined as

$$S_{XY}(\omega) = \lim_{T \to \infty} \frac{1}{2T} X_T^*(\omega) Y_T(\omega) \, , \tag{4.11}$$

then using equations (4.7), (4.8), and (4.3) we obtain

$$S_{XY}(\omega) = S_{XX}(\omega) G(\omega) \, . \tag{4.12}$$

This last equation yields the transfer function for both magnitude and phase in terms of the input- and cross-power spectral densities.

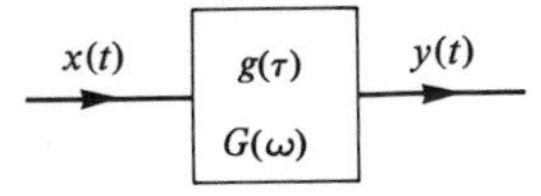

Figure 4.2. Diagram of a SISO system.

4.3 Multiple-input single-output (MISO) systems

Bendat and Piersol derived an expression for the transfer **vector** of a multiple-input single-output (MISO) system (figure 4.3) by a transformation of the system correlation function matrix. Below their result is obtained by an extension of the method outlined in the previous section. For a MISO system it is assumed that it can be characterised by N separate transfer blocks as shown in figure 4.4.

Using the principle of linear superposition and applying the convolution theorem (4.3) we obtain the output of the system of figure 4.4 in the form

$$y_1 = x_1 * g_{11} + x_2 * g_{21} + x_3 * g_{31} + \ldots + x_n * g_{n1} \ . \tag{4.13}$$

Taking the Fourier transform of this equation we find

$$Y_1(\omega) = X_1(\omega)G_{11}(\omega) + X_2(\omega)G_{21}(\omega) + \ldots + X_N(\omega)G_{N1}(\omega) \ . \tag{4.14}$$

Multiplying equation (4.14) by the complex conjugate of $X_1(\omega)$, $X_1^*(\omega)$, and then successively by X_2^*, X_3^* etc. and proceeding to the limit, we obtain the following N simultaneous equations:

$$S_{XY_{11}}(\omega) = S_{XX_{11}}(\omega)G_{11}(\omega) + S_{XX_{12}}(\omega)G_{21}(\omega) + \ldots + S_{XX_{1N}}(\omega)G_{N1}(\omega) \ ,$$

$$S_{XY_{21}}(\omega) = S_{XX_{21}}(\omega)G_{11}(\omega) + S_{XX_{22}}(\omega)G_{21}(\omega) + \ldots + S_{XX_{2N}}(\omega)G_{N1}(\omega) \ ,$$

$$\cdot \quad \cdot \quad \cdot \quad \cdot \quad \cdot \quad \cdot \quad \cdot \quad \cdot \quad \cdot \quad \cdot \quad \cdot \quad \cdot$$

$$S_{XY_{N1}}(\omega) = S_{XX_{N1}}(\omega)G_{11}(\omega) + S_{XX_{N2}}(\omega)G_{21}(\omega) + \ldots + S_{XX_{NN}}(\omega)G_{N1}(\omega) \ .$$

These may be written in matrix notation as

$$S_{XY}(\omega) = S_{XX}(\omega)G(\omega) \ , \tag{4.15}$$

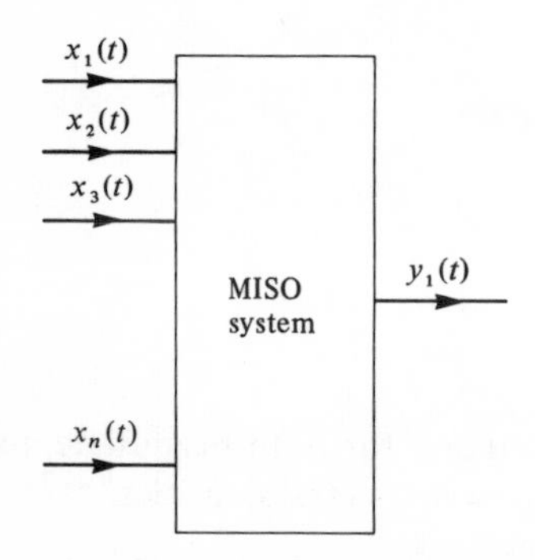

Figure 4.3. Diagram of a MISO system.

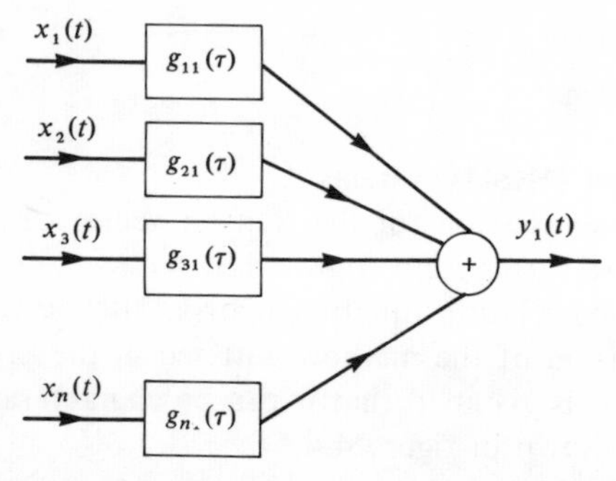

Figure 4.4. Representation of MISO system by N transfer blocks.

where

$$S_{XY}(\omega) = [S_{XY_{11}}(\omega) \;\; S_{XY_{21}}(\omega) \;\; \ldots \;\; S_{XY_{N1}}(\omega)] \,,$$

$$S_{XX}(\omega) = \begin{bmatrix} S_{XX_{11}}(\omega) & S_{XX_{12}}(\omega) & \ldots & S_{XX_{1N}}(\omega) \\ S_{XX_{21}}(\omega) & S_{XX_{22}}(\omega) & \ldots & S_{XX_{2N}}(\omega) \\ \cdot & \cdot & \ldots & \cdot \\ S_{XX_{N1}}(\omega) & S_{XX_{N2}}(\omega) & \ldots & S_{XX_{NN}}(\omega) \end{bmatrix}$$

and

$$G'(\omega) = [G_{11}(\omega) \;\; G_{12}(\omega) \;\; \ldots \;\; G_{1N}(\omega)] \,,$$

where $G'(\omega)$ is the transpose of $G(\omega)$.

Equation (4.15) may be formally solved to yield

$$G(\omega) = S_{XX}^{-1}(\omega) S_{XY}(\omega) \,. \tag{4.16}$$

4.4 Multiple-input multiple-output (MIMO) systems

The above analysis is now extended to systems where there are M outputs as shown in figure 4.5.

Applying the principle of linear super-position and the convolution theorem as before, we derive the relationship between the input and output variables in the form of equations

$$\begin{aligned} y_1 &= x_1 * g_{11} + x_2 * g_{21} + \ldots + x_n * g_{n1} \,, \\ y_2 &= x_1 * g_{12} + x_2 * g_{22} + \ldots + x_n * g_{n2} \,, \\ &\cdot \quad \cdot \quad \cdot \quad \cdot \quad \cdot \quad \cdot \\ y_m &= x_1 * g_{1m} + x_2 * g_{2m} + \ldots + x_n * g_{nm} \,. \end{aligned}$$

Taking the Fourier transform gives

$$\begin{aligned} Y_1 &= X_1 G_{11} + X_2 G_{21} + \ldots + X_N G_{N1} \\ Y_2 &= X_1 G_{12} + X_2 G_{22} + \ldots + X_N G_{N2} \\ &\cdot \quad \cdot \quad \cdot \quad \cdot \quad \cdot \quad \cdot \\ Y_M &= X_1 G_{1M} + X_2 G_{2M} + \ldots + X_N G_{NM} \,, \end{aligned} \tag{4.17}$$

in which the frequency dependence has been omitted for clarity.

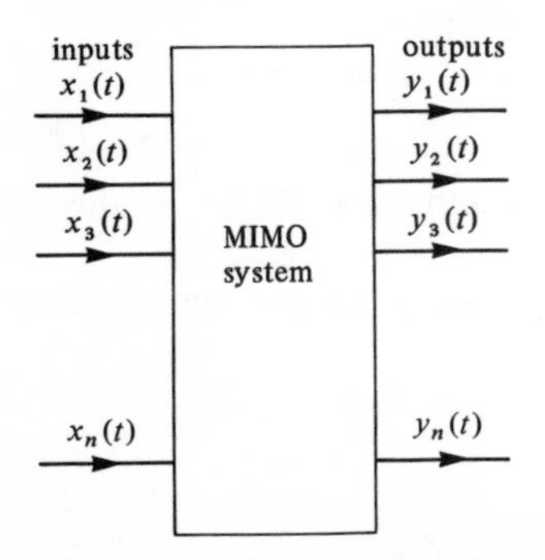

Figure 4.5. Diagram of a MIMO system.

Equation (4.17) can be written in the matrix form

$$[Y_1, Y_2, ..., Y_M] = [X_1, X_2, ..., X_N]\begin{bmatrix} G_{11} & G_{12} & \dots & G_{1M} \\ G_{21} & G_{22} & \dots & G_{2M} \\ \cdot & \cdot & \dots & \cdot \\ G_{N1} & G_{N2} & \dots & G_{NM} \end{bmatrix}$$

or simply

$$Y = XG \,. \tag{4.18}$$

Premultiplying equation (4.18) by a column vector of the complex conjugates X_1^*, X_2^* etc. we obtain

$$X^*Y = X^*XG \,.$$

Using the definitions of equations (4.8) and (4.11) we can now write

$$\underset{N\times M}{[S_{XY}]} = \underset{N\times N}{[S_{XX}]}\underset{N\times M}{[\ G\ ]} \,,$$

where $N \times M$ means a matrix of N rows and M columns, or simply

$$S_{XY} = S_{XX}G \,. \tag{4.19}$$

Equation (4.19) may be formally solved to yield

$$G(\omega) = S_{XX}^{-1}(\omega)S_{XY}(\omega) \,. \tag{4.20}$$

Equation (4.20) is completely general; it will be shown in succeeding sections how the power spectral densities can be used to measure the transfer function relationships.

4.5 Summary

We have seen that in all three cases of SISO, MISO, and MIMO systems the final equation for the system is identical and has the form

$$S_{XY}(\omega) = S_{XX}(\omega)G(\omega) \,. \tag{4.21}$$

This means that a measurement of the input P.S.D. and the cross-spectrum between input and output, followed by a simple computation, yields the transfer function of the system under investigation.

Taking Fourier transforms of equation (4.21) we have

$$R_{xy}(\lambda) = R_{xx}(\lambda) * g(\lambda) \,, \tag{4.22}$$

where $R_{xx}(\lambda)$ is the autocorrelation of the input and $R_{xy}(\lambda)$ is the cross-correlation between input and output.

If the input signal is 'white', i.e. with equal power per unit bandwidth, then we have

$$S_{XX}(\omega) = A \,,$$

a constant, say, and

$$R_{xx}(\lambda) = A\delta(\lambda)$$

where $\delta(\lambda)$ is a delta function. Under these conditions we have

$$G(\omega) = \frac{1}{A} S_{XY}(\omega) ,$$

$$g(\lambda) = \frac{1}{A} R_{xy}(\lambda) . \tag{4.23}$$

In other words, these last two equations—which say the same thing in different ways—emphasise that, if an input signal is white, its autocorrelation function is a delta function, and its transfer function and impulse response are proportional to the cross-spectrum and the cross-correlation respectively.

4.6 Coherence functions

Returning once again to the simple case of a single-input single-output (SISO), if the system is linear and the input $x(t)$ and output $y(t)$ have Fourier transforms $X(\omega)$ and $Y(\omega)$ respectively, then

$$Y(\omega) = X(\omega)G(\omega) . \tag{4.24}$$

Proceeding as before we obtain

$$G(\omega) = \frac{S_{XY}(\omega)}{S_{XX}(\omega)} . \tag{4.25}$$

Alternatively, we have

$$|G(\omega)|^2 = \frac{S_{YY}(\omega)}{S_{XX}(\omega)} . \tag{4.26}$$

Taking the modulus of equation (4.25) and dividing out by equation (4.26) we obtain

$$G_{XY}(\omega) = \frac{|S_{XY}(\omega)|^2}{S_{XX}(\omega)S_{YY}(\omega)} = 1 \tag{4.27}$$

in this case. $G_{XY}(\omega)$ is defined as the coherence function and measures the degree of linear dependence between a pair of signals. In a linear system then the coherence function is unity. In general, however, we have

$$G_{XY}(\omega) = \frac{|S_{XY}(\omega)|^2}{S_{XX}(\omega)S_{YY}(\omega)} \leqslant 1 . \tag{4.28}$$

4.7 Pseudo-random binary signals

A number of applications are now being carried out using pseudo-random binary signals as the input test signal. The autocorrelation function of an infinite length of a random binary signal consists of a delta function at the origin and zero elsewhere. Binary signals which are random over a finite length and then repeat are called pseudo-random binary (PRBS), and have for an autocorrelation a series of functions which approximate to delta functions spaced at the period of repeatability of the sequence, as shown in figure 4.6.

A suitable PRBS signal can be generated by use of a number of shift register stages n and suitable feedback connections. A 127 digit PRBS generator is shown in the block diagram of figure 4.7. Other length sequences can be constructed in a similar manner, as illustrated in table 4.1.

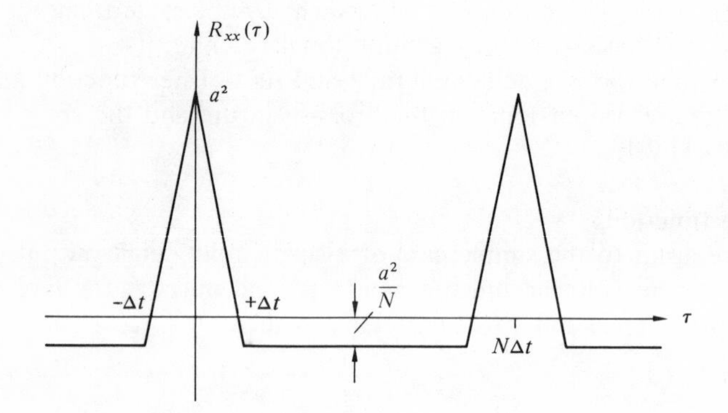

Figure 4.6. Autocorrelation function of a PRBS with period $N\Delta t$ and levels $+a$; $-a$.

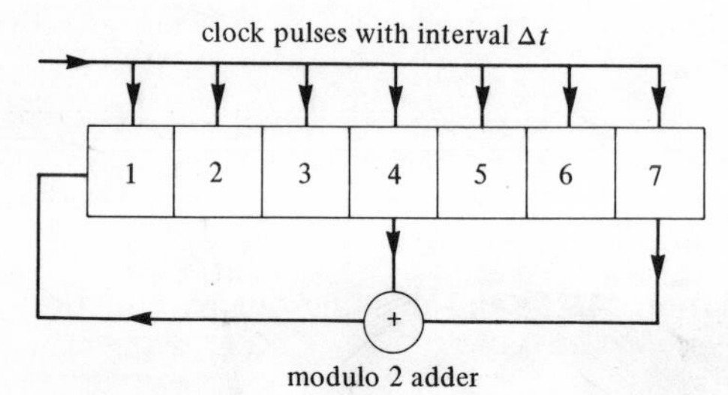

Figure 4.7. Block diagram of a PRBS generator.

Table 4.1. Pertinent data for PRBS generator for different numbers of shift registers.

Number of shift registers n	Period of sequence $N(= 2^n - 1)$	Feedback to 1st stage—the mod 2 sum of output of stages
2	3	1 and 2
3	7	2 and 3
4	15	3 and 4
5	31	3 and 5
6	63	5 and 6
7	127	4 and 7
8	255	2, 3, 4 and 8
9	511	5 and 9
10	1023	7 and 10
11	2047	9 and 11

5

Use of noise analysis in measurement

5.1 Introduction
The various techniques of noise analysis are usually employed to find out some details concerning either a source of noise or a system through which a signal has passed. More than one of the techniques described in earlier chapters will probably be employed; however, initially we shall consider what inferences can be drawn from each technique employed on its own.

5.2 Uses of autocorrelation
It has been said that there was only one person who could effectively interpret an autocorrelogram, and he is now dead. However, the autocorrelogram as such still has some uses, and one of the most important is its use in determining whether or not there is a periodic signal buried in noise. The autocorrelogram can provide a form of signal recovery as illustrated by the sketches of figure 5.1.

The value of the ordinate of the autocorrelation for zero time lag gives the variance for a signal having a zero mean. The square root of this variance is the *true* r.m.s. (root mean square) value for the signal.

The autocorrelogram for a single degree of freedom system excited by white noise is a decaying exponential cosine wave, and the periodicity

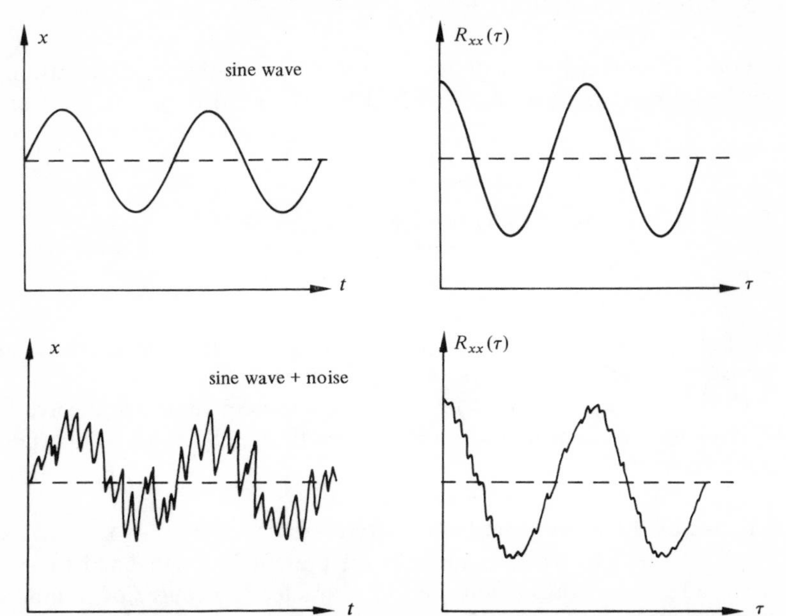

Figure 5.1. Illustration of recovery of periodic signal using autocorrelation.

and damping coefficient of the autocorrelogram are the same as for free oscillation of the original system.

To prove the previous statement let us consider a single degree of freedom system described by the equation

$$\frac{d^2x}{dt^2}+2b\frac{dx}{dt}+\omega_0^2 x = \frac{F}{m}\,, \tag{5.1}$$

where x is displacement, t is time, b is a damping coefficient, ω_0 the natural angular frequency of the system with no damping, m is the mass, and F is the applied force. The solution to equation (5.1) for free oscillation ($F = 0$) is

$$x = A\,e^{-bt}\cos(\omega' t+\phi)\,, \tag{5.2}$$

where $\omega'^2 = \omega_0^2 - b^2$, and A and ϕ are amplitude and phase factors respectively, which depend on the initial conditions.

The transfer function of a system described by equation (5.1) is given in the frequency domain by

$$H(j\omega) = \frac{1}{-\omega^2+2jb\omega+\omega_0^2}\,. \tag{5.3}$$

If the force has a white noise characteristic, then $S_{FF}(\omega) = K$, and thus the P.S.D. of the system, is given by:

$$S_{xx}(\omega) = S_{FF}(\omega)|H(\omega)|^2 = \frac{K}{(\omega_0^2-\omega^2)+4b^2\omega^2}\,. \tag{5.4}$$

A Fourier transformation of equation (5.4) yields the autocorrelation for the response of the system (Wiener–Khintchine relationship), and this is

$$R_{xx}(\tau) = \frac{\pi K}{4bm}e^{-b|\tau|}\left(\cos\omega'\tau+\frac{b}{\omega'}\sin\omega'|\tau|\right)\,. \tag{5.5}$$

If $b/\omega' \ll 1$, equation (5.5) becomes:

$$R_{xx}(\tau) \approx \frac{\pi K}{4bm}e^{-b|\tau|}\cos\omega'\tau\,. \tag{5.6}$$

Equation (5.6) shows the same periodicity and decay characteristics as the displacement in the basic system [equation (5.2)].

Thus we have a means of investigating a system via the autocorrelogram without having to sustain a resonance condition. Such an autocorrelogram is shown in figure 5.2.

5.3 Uses of the power spectral density (P.S.D.)

The P.S.D. tells us what frequencies are present in a signal and their magnitude, and is thus often used to characterise a source of sound. This aspect is often known as character recognition.

Frequency analysis of signals is most useful in studying vibration phenomena (any discrete frequencies correspond to mechanical resonances), speech, acoustic noise, turbulence, and sonar signals. The frequency and magnitude of discrete frequencies arising from a vibrating body can often be used to isolate some malfunction by identifying these frequencies with various sources of vibration.

One method of determining the response of a system frequency-wise is to use a sine wave input whose frequency can be varied and then the input and output are compared for different frequencies at the input so as to obtain the response. If a noise signal (or some form of complex signal) is used as an input, then the system is being simultaneously excited by a wide range of frequencies and thus an analysis of the input and output using the P.S.D. will also allow us to determine the response for the range of frequencies present in the signals (except when there is additive noise in the system). A noise signal may have been used as an input because it was not possible to achieve a sine wave input under the experimental conditions, or the use of a sine wave input may excite some resonant modes which could disturb the parameter being measured. With reference to this last point, the measurement of the acoustical properties of rooms reveals that certain standing waves can be excited in the room, if an acoustic source at a single frequency corresponding to a room mode is employed for any sustained length of time. In normal speech conditions it is unlikely that a single-frequency signal will persist for enough time to establish one of these room modes, and thus the acoustic signal should be chosen bearing these facts in mind. The use of noise as a test signal means that we have a randomness in the phase relations for each frequency present in the noise signal (in contradistinction to the coherent nature of a sine wave) and thus there is no possibility of a normal mode being excited.

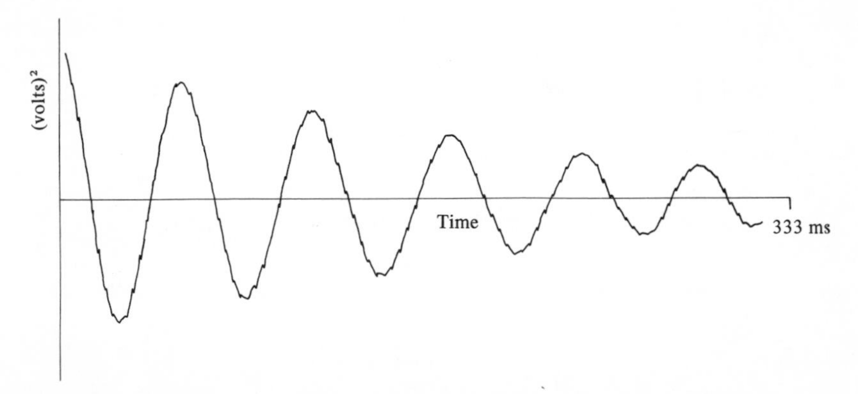

Figure 5.2. Autocorrelogram for a single degree of freedom system excited by white noise.

The use of the P.S.D. in both the input and output signals, so as to determine the transfer function of a system, is severely restricted if the observed input and output are in reality the true input and output plus noise. The most general situation that can occur is depicted in figure 5.3. The noise source at the input $u(t)$ could arise from the monitoring equipment, and the noise source at the output $v(t)$ could arise from the monitoring equipment and/or noise sources within the system.

The determination of the transfer function (response) of a system where the measurements are typified by figure 5.3 can be carried out using the cross-spectrum, and this procedure will be detailed in section 5.5 where the uses of cross-correlation are discussed.

An important use of the P.S.D. is to determine the effective loudness of an acoustic noise as perceived by the ear, and this employs analysis in $\frac{1}{3}$-octave bands as detailed in section 3.1.3. The weighted analyses (to

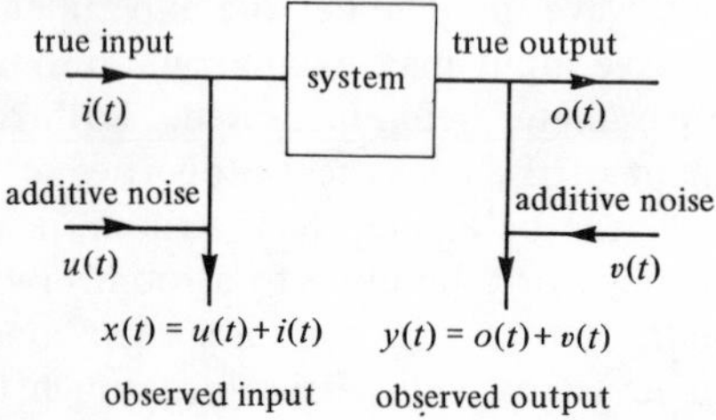

Figure 5.3. Effect of noise on the measurement of the parameters of a system.

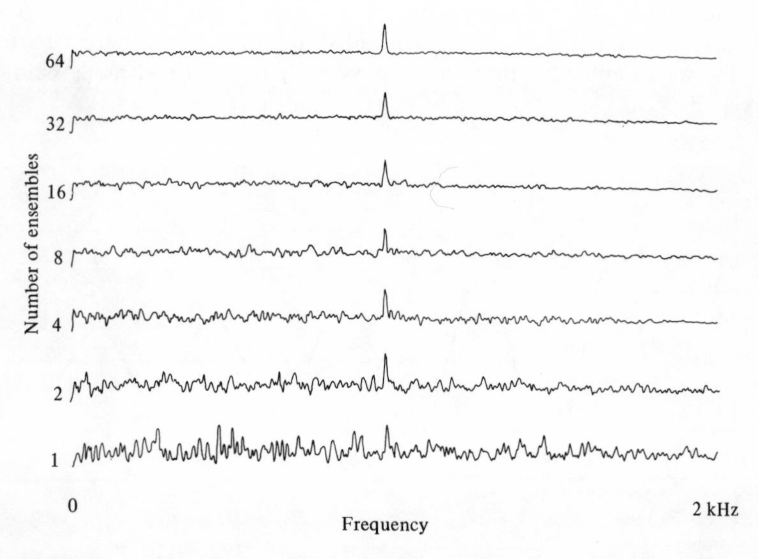

Figure 5.4. Effect of ensemble averaging on the P.S.D. of a sine wave buried in noise.

compensate for ear characteristics) on a logarithmic scale are referred to as phons.

Some improvement can be obtained in the resolution of the P.S.D. by averaging many consecutive spectra. This is referred to as ensemble averaging and its effects are illustrated in figure 5.4 where a signal containing a sine wave buried in noise is being analysed, the sine wave being 30 dB below the level of the noise.

5.4 Uses of probability distribution function (p.d.f.)

The p.d.f. is a statistical function and allows us to determine the probability that a given signal level should occur, and, by integration, the probability that signals greater than or less than a given level may occur. Thus some idea of the extent of the excursions of the signal levels and the probability of these excursions is available from the p.d.f. The form of the p.d.f. curve can also give us some idea of the nature of the signal

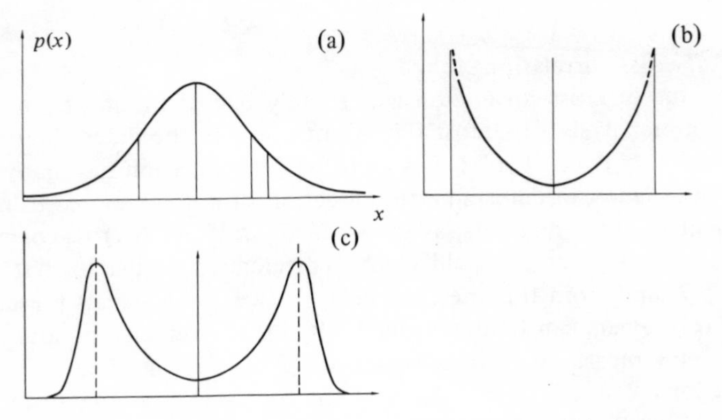

Figure 5.5. Examples of p.d.f.: (a) random noise; (b) sine wave; (c) random noise + sine wave.

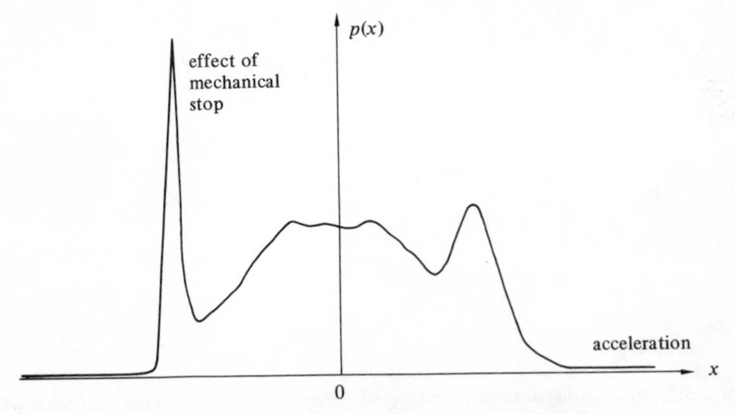

Figure 5.6. P.d.f. for vibration constrained by a mechanical stop.

being analysed, as exemplified in figure 5.5. Actually the p.d.f. of figure 5.5(c) will only be observed when the sine wave is of about the same level as the noise. A random noise signal will often show a Gaussian distribution and thus we can state that we do have a random noise input if we observe this Gaussian distribution. One must be careful not to fall into the trap of trying to make all random noise data fit a Gaussian p.d.f. curve, even though this is usually the case. A Gaussian p.d.f. has also a Gaussian characteristic function, as the Fourier transform of a Gaussian curve is also a Gaussian curve.

It is possible that another function, the peak p.d.f. (which tells what proportion of time is spent by the peaks of the signals at various amplitudes), and the characteristic function may be useful parameters when considering the analysis of signals passing through nonlinear systems (see chapter 7). The p.d.f. of the flow-induced vibration in a mechanical system constrained by a mechanical stop is shown in figure 5.6.

5.5 Uses of cross-correlation

The technique of cross-correlation is definitely one of the most powerful tools in noise analysis. Its most significant use is in the determination of the time delay between two signals and this determination has many applications. One can determine the direction of a noise source using an arrangement of three receivers, as shown in figure 5.7. A cross-correlation between receivers *1* and *2* would yield a correlogram similar to that shown in figure 5.7, and from this the time delay between the signals presented to the two receivers can be determined. If v is the velocity of sound, τ_2 the time delay measured for receivers *1* and *2*, and τ_3 the time delay measured for receivers *1* and *3*, then, if the source is at a large distance

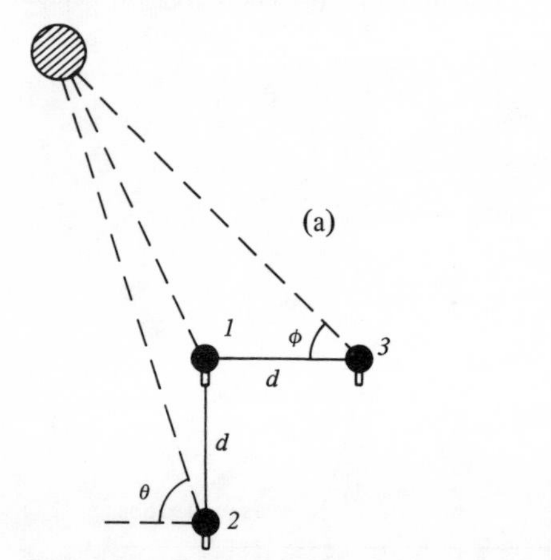

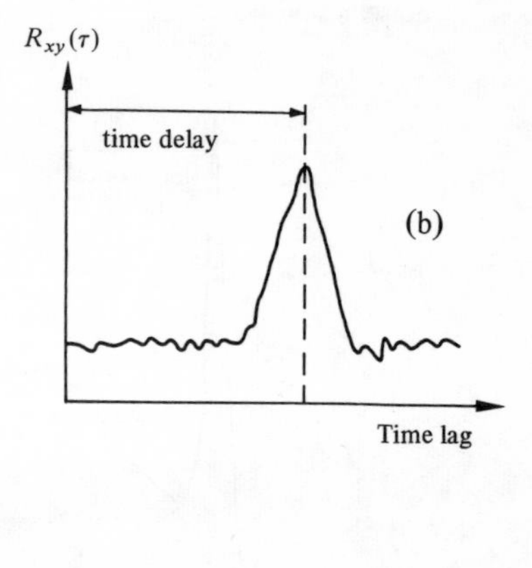

Figure 5.7. Determination of direction of noise source. (a) Experimental arrangement of three transducers; (b) typical cross-correlogram.

compared with the distance d between the receivers, simple trigonometry shows that

$$\sin\theta = \frac{v\tau_2}{d}\ , \qquad \cos\phi = \frac{d}{v\tau_3}\ . \tag{5.7}$$

The measurement of acoustic absorption coefficients can be carried out *in situ* using the arrangement shown in figure 5.8, where the cross-correlation between the noise source and the microphone is also shown. The amplitude ratio of the cross-correlation peaks corresponding to the direct path *1* and the path after reflection from the panel under test *2* yields a measure of the acoustic absorption of the panel.

The velocity of a white hot steel strip can be determined using a cross-correlation between the output from two photocells as shown in figure 5.9. The noise output from the photocells is due to irregularities

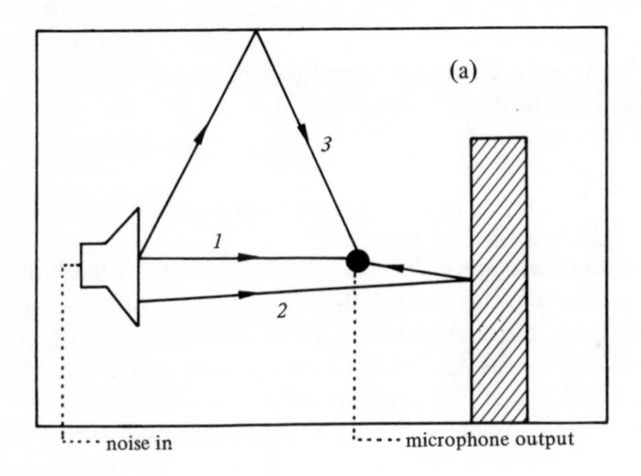

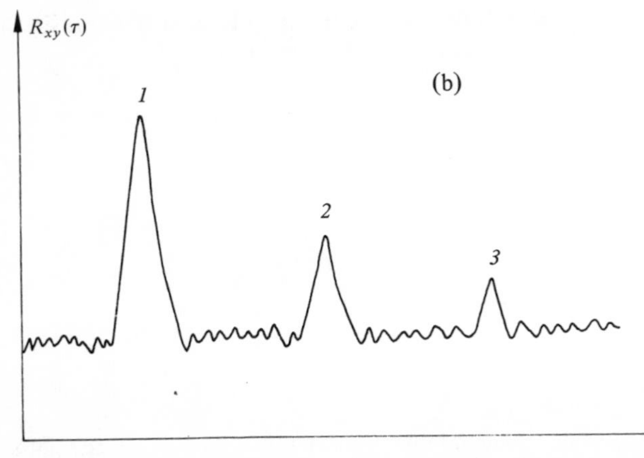

Figure 5.8. Determination of acoustic absorption *in situ.* (a) Experimental arrangement; (b) cross-correlogram between noise source and microphone.

in the metal surface and the maximum in the cross-correlation corresponds to the time taken for these irregularities to traverse the distance between the two sensors.

The velocity of bubbles in a fluid can be determined by cross-correlating the outputs of two conductivity probes separated by a set distance in the bubbly fluid.

It is not possible to draw up an exhaustive list of the applications of cross-correlation, as the possible applications are only limited by the ingenuity of the researcher. A word of caution must be sounded here about using cross-correlation in systems exhibiting dispersion (velocity of propagation of a disturbance is a function of frequency) and systems exhibiting an oscillation having a random amplitude. In attempting to cross-correlate signals from such systems a null result often occurs due to frequency shifting as a result of the dispersion or distortion of the random amplitude oscillation (due to change in frequency components for side bands of an amplitude modulated signal). The effect of a frequency shift means that what was $A\sin\omega t$ at one transducer becomes $B\sin[(\omega+\delta\omega)t+\tau]$ at the other transducer. The contribution to the cross-correlation at lag τ is now

$$R_{xx}(\tau) = \lim_{T\to\infty}\frac{1}{T}\int_0^T A\sin(\omega t+\tau)B\sin[(\omega+\delta\omega)t+\tau]\,\mathrm{d}t\,. \tag{5.8}$$

Equation (5.5) is now zero since $\omega \neq \omega+\delta\omega$. However, if we look at the cross-correlation of the envelopes of (filtered) signals we can by-pass this phenomenon and get a meaningful result. This is illustrated in figure 5.10 where a cross-correlation was carried out between two accelerometers situated on a long rod along which a mechanical impulse was being propagated. The cross-correlation on the two raw signals [figure 5.10(a)] showed no correlation whereas the cross-correlation between the envelopes of the signals showed a definite peak corresponding to the transit time for mechanical waves between the two accelerometers.

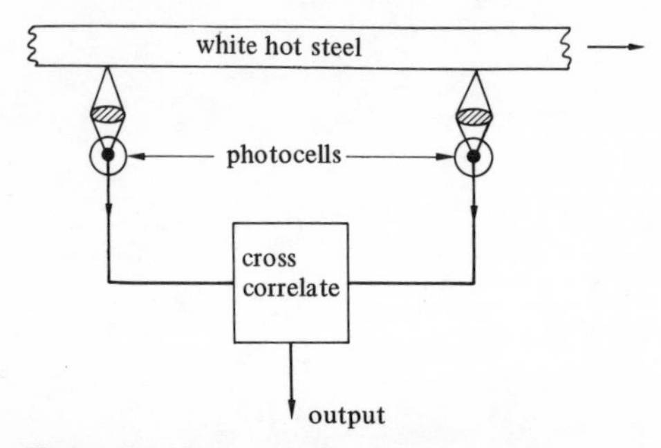

Figure 5.9. Arrangement for the measurement of the velocity of a white hot steel plate.

An important quantity derived from the cross-correlation is the cross-spectrum, which is simply the Fourier transform of the cross-correlation. The cosine transformation is usually referred to as the co-spectrum and the sine transformation as the quad-spectrum. Let us now consider again the system with additive noise shown in figure 5.3. We have already stated that a determination of the transfer function using a ratio of the P.S.D. of output and input will be in error if there is additive noise. First, let us consider the case where there is additive noise only at the output $[u(t) = 0, v(t) \neq 0]$. Now, this noise $v(t)$ bears no relation to the input, and thus when we perform a cross-correlation calculation (and then cross-spectrum calculation) between input and output there will be no contribution due to additive noise. Hence, if S_{yx} is the cross-spectral density between y and x [note that $x(t) = i(t)$], and S_{xx} is the power spectral density of x, then, if $H(f)$ is the transfer function of the system (f being frequency), we have

$$S_{yx}(f) = H(f)S_{xx}(f) \,. \tag{5.9}$$

Also, we can obtain information about the noise spectral density, $S_{vv}(f)$, using equation (5.7):

$$S_{yy}(f) = |H(f)|^2 S_{xx}(f) + S_{vv}(f) \,. \tag{5.10}$$

For a system where additive noise is present at both input and output there is no simple way of obtaining the transfer function. Various workers are currently searching for means of providing a reliable estimate of $H(f)$, and we shall mention here only one technique of limited application. If a

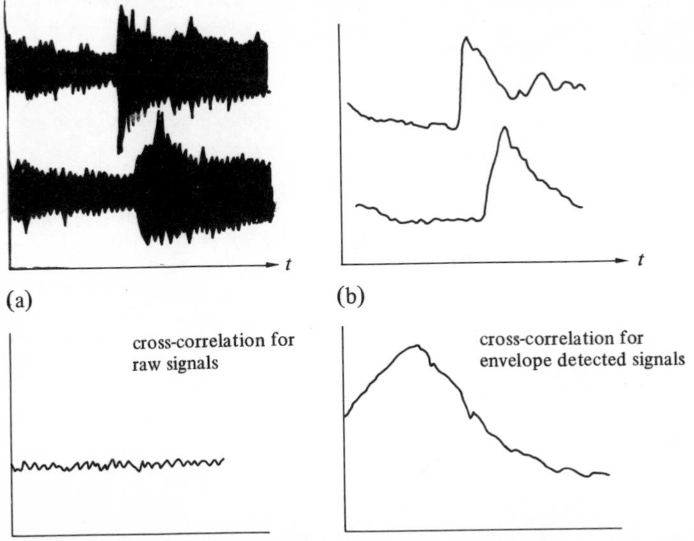

Figure 5.10. Cross-correlograms for mechanical waves along a rod. (a) Raw signal; (b) envelope detected signals.

signal $j(t)$ is available, which is correlated to both $x(t)$ and $y(t)$ and also statistically independent of $u(t)$ and $v(t)$, then the cross-correlation between $x(t)$ and $j(t)$ is independent of the additive noise $u(t)$, and similarly the cross-correlation between $y(t)$ and $j(t)$ is independent of the additive noise $v(t)$. Hence, in terms of cross-spectral densities we have

$$S_{yj}(f) = H(f)S_{xj}(f)\,. \tag{5.11}$$

It will be noticed from equation (5.6) that, since $S_{yx}(f)$ will be in general complex, so will $H(f)$, and thus we can gain information on both the amplitude and phase characteristics of the transfer function, whereas the ratio of output/(input P.S.D.) gives only the modulus of the transfer function as shown in equation (5.9) (if there is no additive noise)

$$|H(f)|^2 = \frac{S_{yy}(f)}{S_{xx}(f)}\,. \tag{5.12}$$

Analysis of errors and confidence limits

6.1 Some statistics

The three most important statistical parameters often used to characterise a group of variables are the mean, variance [which is $R_{xx}(0)$], and standard deviation [= (variance)$^{1/2}$]. The mean gives a measure of the most expected value of the variable, and the standard deviation gives a measure of the spread of values about the mean. A statistical distribution gives a means of predicting the probability that a variable lies between certain values. A most important distribution is the normal or Gaussian distribution as random signals generally have the form of this distribution for their probability density functions. The normal distribution is of the form:

$$p(x) = \frac{1}{\sigma(2\pi)^{1/2}} \exp\left[-\frac{(x-\mu)^2}{2\sigma^2}\right], \tag{6.1}$$

where μ is the mean and σ the standard deviation. The probability that the variable should have a value of x in the range x to $x+\delta x$ is $p(x)\delta x$. Figure 6.1 shows the form of a Gaussian distribution and it will be seen that the distribution is symmetrical and extends either side of the mean by about 3σ. It turns out that about 68% of the values lie between $\pm\sigma$, and about 95% of the values lie between $\pm 2\sigma$. Thus, if we look $\pm 2\sigma$ on either side of the mean, we have a 95% confidence on finding the value of the variable. Another important property of the Gaussian distribution is that its Fourier transform (i.e. the characteristic function) is itself a Gaussian distribution.

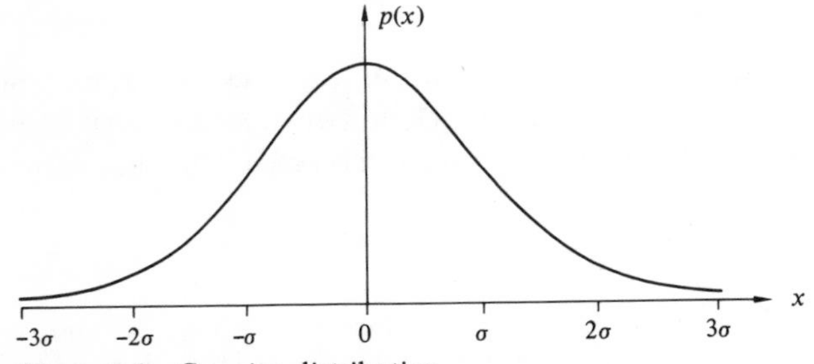

Figure 6.1. Gaussian distribution.

6.2 Errors in correlation

If we determine the cross-correlation for a sample of time T, then the value of the cross-correlation will vary for different samples. If we let $x(t)$ and $y(t)$ be the two signals, $R_{xy}(\tau, T)$ be the cross-correlation for a sample of length of time T, and let $R_{xy}(\tau)$ be the cross-correlation which is the average of many values of $R_{xy}(\tau, T)$ (which is the same as for $T \to \infty$), the ensemble average, then a measure of the error in the cross-correlation is the variance $\sigma_{xy}^2(\tau, T)$, defined by

$$\sigma_{xy}^2(\tau, T) = \langle [R_{xy}(\tau, T) - R_{xy}(\tau)]^2 \rangle_{av} . \tag{6.2}$$

$$= \langle R_{xy}^2(\tau, T) \rangle_{av} - 2\langle R_{xy}(\tau, T) R_{xy}(\tau) \rangle_{av} + R_{xy}^2(\tau)$$

$$= \langle R_{xy}^2(\tau, T) \rangle_{av} - R_{xy}^2(\tau) . \tag{6.2'}$$

Now, if we look closely at the first term on the right hand side of equation (6.2′), we see that it is given by

$$\langle R_{xy}^2(\tau, T) \rangle_{av} = \frac{1}{T^2} \int_0^T \int_0^T \langle x(t) y(t+\tau) x(u) y(u+\tau) \rangle_{av} \mathrm{d}t \, \mathrm{d}u . \tag{6.3}$$

Expression (6.3) depends on the fourth-order distribution moment for the random process $P_{xy}^2(\tau, t, u)$, defined as

$$P_{xy}^2(\tau, t, u) = \langle x(t) y(t+\tau) x(u) y(u+\tau) \rangle_{av} . \tag{6.4}$$

Expression (6.4) involves quite a formidable calculation. However, when considering the autocorrelation of a Gaussian random noise process, we obtain a simpler expression for the fourth-order distribution moment:

$$P_{xx}^2(\tau, v) = R_{xx}^2(\tau) + R_{xx}^2(v) + R_{xx}(\tau + v) R_{xx}(\tau - v) , \tag{6.5}$$

where $v = u - t$. For large values of T, equation (6.2′) can be shown to be

$$\sigma_{xy}^2(\tau, T) \rightarrow \frac{2}{T} \int_0^T [P_{xy}^2(\tau, v) - R_{xy}^2(\tau)] \, \mathrm{d}v . \tag{6.6}$$

The output signal to noise amplitude ratio is given by

$$\frac{\text{expected mean value}}{\text{standard deviation}} = \frac{R_{xy}(\tau)}{\sigma_{xy}(\tau, T)} . \tag{6.7}$$

If we are dealing with a normal distribution, then we have already seen that for a 95% certainty we have to include two standard deviations, so that, for a 95% certainty that an arbitrary measured value lies within p% of a true mean, we can put

$$0{\cdot}01 p R_{xy}(\tau) = 2\sigma_{xy}(\tau, T) . \tag{6.8}$$

Equation (6.8) allows us to find the relationship between the finite integration time T and the correlation displacement (time lag) in order to guarantee a desired accuracy of a measurement along the correlation curve.

If we consider a signal passing through a filter, then the expression for $\sigma_{xy}(\tau, T)$ involves the product BT (where B is the bandwidth), and, to achieve our 95% certainty that a measured value lies within less than 20% of the true mean value, the BT product must be at least 100. Thus, small bandwidths require large sample times and vice versa. We have in fact a form of uncertainty principle

$$BT \geqslant 100 \tag{6.9}$$

and any violation of this condition will lead to statistical unreliability in one of the quantities.

6.3 Errors in determination of frequency spectra

If we define the number of degrees of freedom for a spectral analysis to be $2T\Delta f$, where T is the duration of the signal being analysed and Δf the bandwidth, then we can compute a curve showing the possible deviations from the true value of the spectral density versus the number of degrees of freedom for various confidence limits. This curve is shown in figure 6.2.

We have seen in section 6.2 that there exists a bandwidth–time product uncertainty relation, and we see that this also applies to analysis in the frequency domain. We can look at this aspect in two other ways which will make it clearer how serious this limitation is. If we perform a frequency analysis on a segment of length T of a continuous sine wave then we do not get a line frequency but a function of the form $[(\sin\psi)/\psi]^2$ centred at the frequency of the sine wave, say f_0. This is

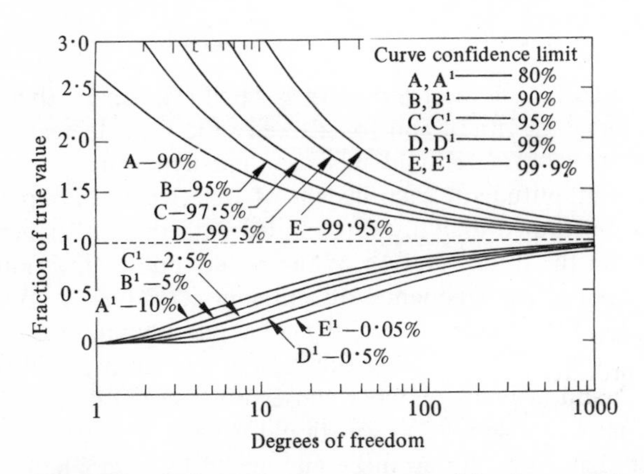

Figure 6.2. Relation between degrees of freedom and fraction of true values for several values of confidence limits for power-spectral-density measurements (from W. E. Schiesser, *Statistical Uncertainty of Power Spectral Density Estimates*, Bulletin 711-C1, Weston-Boonshaft & Fuchs, 1966).

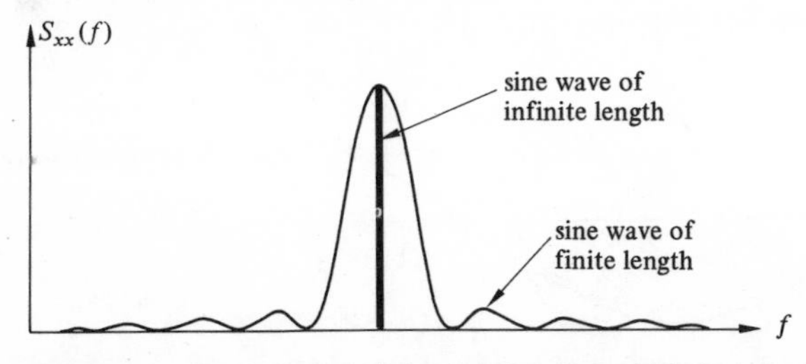

Figure 6.3. Power spectral density for truncated sine wave.

illustrated in figure 6.3. As $T \rightarrow \infty$ the spectrum tends to a line spectrum, so for extremely small bandwidths (good resolution) we require a long sample.

Again, we can see that a precise determination of frequency requires a long time base (T) and a precise determination of time for a fixed time base [and hence determination of $x(t)$ at a given time] requires a broad bandwidth, by noting that (the proof will not be reproduced here)

$$T_0 \Omega_0 > 1/4\pi \,, \tag{6.10}$$

where T_0 and Ω_0 are measures of the spread of the function $x(t)$ about the origin and the spread of the Fourier transform of $x(t)$, $A(f)$, on the frequency scale. T_0 and Ω_0 are defined by the following equations

$$\left.\begin{aligned} T_0 &= \left[\int_{-\infty}^{\infty} t^2 x(t)\,\mathrm{d}t\right]^{1/2} \\ \Omega_0 &= \left[\int_{-\infty}^{\infty} f^2 A(f)^2 \mathrm{d}f\right]^{1/2} \\ A(f) &= \left[\int_{-\infty}^{\infty} x(t)\mathrm{e}^{-\mathrm{i}2\pi f t}\,\mathrm{d}t\right]^{1/2} \end{aligned}\right\} . \tag{6.11}$$

Thus, from equation (6.10), if the bandwidth is small (Ω_0 small) then the time base must be large (T_0 large), and for a fixed time base decreasing the bandwidth requires that T_0 should increase, and hence the determination of the amplitude at each instant of time becomes imprecise.

If the analysis is performed digitally on sampled data then a further limitation is placed on the interpretation of the results due to the finite time between samples τ. The frequency of sampling f_s ($= 1/\tau$) is often referred to as the Nyquist frequency and, if f_{max} is the maximum frequency for which the spectral density function is significant, then we must have $f_s > f_{max}$, and in fact—taking a conservative view of the situation (as mentioned in chapter 3)—we should have $f_s > 4f_{max}$. Frequencies greater than f_s should again be eliminated by filters before digital processing commences, otherwise there exists the problem of aliasing. This is illustrated in figure 6.4 where the effect of slow sampling of a fast sine wave produces ordinates that appear to come from a sine wave at a lower frequency. Aliasing can be, and in fact should be, prevented by the use of low-pass filters.

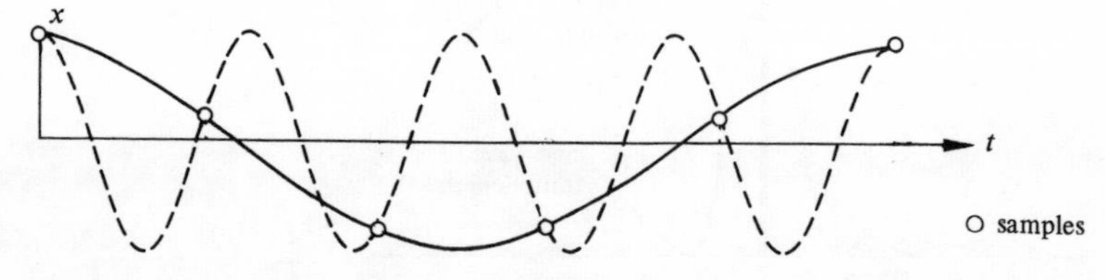

Figure 6.4. Effect of slow sampling of a fast sine wave (aliasing).

The approximation of a signal by discrete ordinates produces errors in itself and a discussion of this was given in chapter 3.

6.4 Effects of truncating a spectral analysis

The analysis of a signal in the frequency domain requires an infinite number of coefficients to represent the contribution at each frequency; however, it is usual to terminate the analysis after a certain finite number of coefficients has been computed (truncate the analysis). We would certainly try to calculate enough coefficients so as to achieve a negligible error due to the neglected terms, but for digital sampling we are limited by the Nyquist frequency in the maximum number of coefficients that we may calculate. If the coefficients are not apparently converging rapidly enough then they are often weighted or smoothed to achieve a better convergency (this is a consequence of 'window carpentry' as described in chapter 3). The effects of truncation are most evident for a signal with sharp discontinuities, and figure 6.5 shows the effect when a square pulse is reconstructed from a truncated spectrum analysis, and also the reconstruction from a smoothed series. If there are no sharp discontinuities, then the errors due to truncation are not as severe as those depicted in figure 6.5.

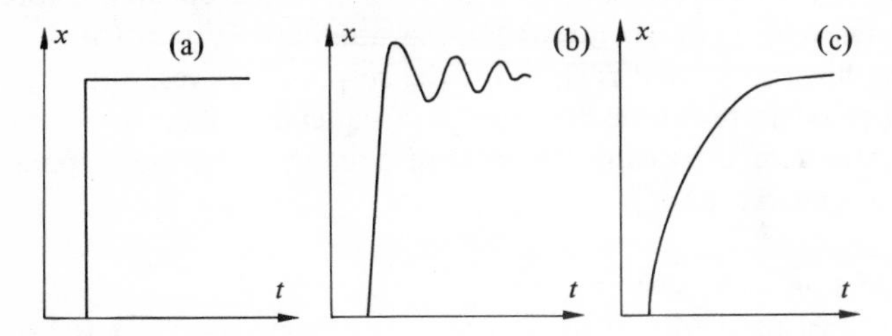

Figure 6.5. Effect of truncation on the reconstruction of a pulse. (a) Original square pulse; (b) reconstruction from truncated spectral analysis; (c) reconstruction from smoothed truncated spectral analysis.

7

Nonstationary and nonlinear systems

7.1 Introduction
The analysis in the preceding chapters has been restricted to stationary and linear systems. A linear system implies that there exists a relationship of the form $y = cx$ between input and output, where c is a constant which may be complex (has a real and imaginary part) and may be a function of frequency. Nonstationary processes may be analysed by considering them to be stationary over small time intervals, and the ramifications of this approach, together with the subject of 'evolutionary' spectra, are treated in section 7.3. Nonlinear processes are much more difficult to analyse and the determination of the system's characteristics from measurements on input and output signals is a well nigh impossible task. However, measurements using the p.d.f. can provide some insight into the system, and this is considered in section 7.3. To give some idea of the problems arising when we are dealing with a nonlinear system, figure 7.1 shows the unusual form of the response curves (amplitude of response versus frequency) for a given level of excitation for a mass-spring-dashpot system, where the spring is linear, hardening (stiffness increasing with increasing displacement) and softening (stiffness decreasing with increasing displacement). Note how the response can undergo jumps for the nonlinear system for frequencies within the shaded regions on the response curves, and thus we can have two possible responses for a single frequency. Also in figure 7.1 is shown the effect of increasing the level of excitation on the response curve for a nonlinear system.

7.2 Analysis of nonlinear systems
Unfortunately measurements of input and output signals cannot tell us a lot about a nonlinear system. It is interesting to see, however, what techniques are employed to analyse nonlinear systems if their characteristics are known. A system is usually characterised by a differential-integral equation. For example, if we consider an inductor L, resistor R, and capacitor C connected in series to a voltage V, then the current i may be found by solving the equation

$$L\frac{\mathrm{d}i}{\mathrm{d}t}+Ri+\left(\frac{1}{C}\right)\int i\,\mathrm{d}t = V\,. \tag{7.1}$$

Equation (7.1) is derived by equating the applied voltage to the sum of potential drops across the components of the system. It can be solved directly, or by means of a transform technique which replaces the differentials and integrals by quantities which can be algebraically manipulated. One such transform is the Laplace transform, $\mathrm{F}(p)$, which is defined for a function of t, $\mathrm{f}(t)$, as

$$\mathrm{F}(p) = \int_0^\infty \mathrm{f}(t)\mathrm{e}^{-pt}\mathrm{d}t\,. \tag{7.2}$$

No restrictions are placed on the parameter p, which can be complex. If a bar is used to denote a transformed variable, the application of the Laplace transform to equation (6.1) yields

$$Lp\bar{i} + R\bar{i} + \frac{1}{Cp}\bar{i} = \bar{V}\,, \tag{7.3}$$

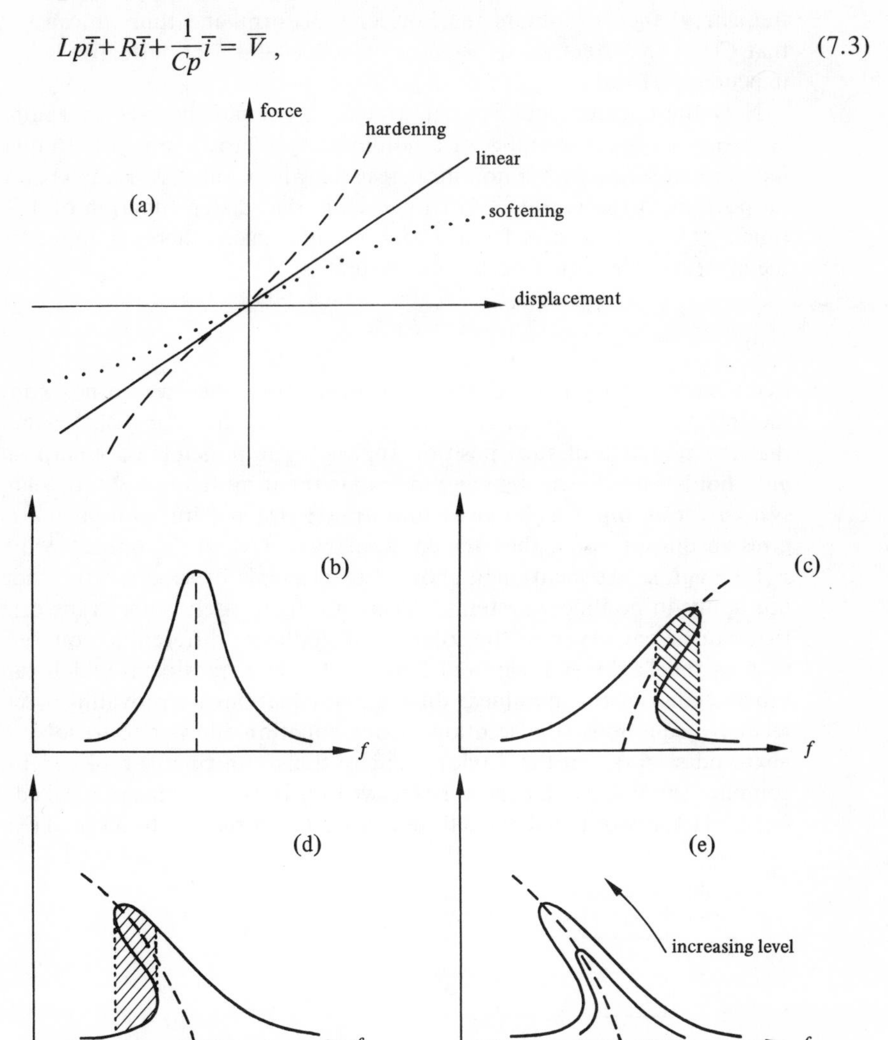

Figure 7.1. Responses of linear, hardening and softening spring systems. (a) Force-displacement curves for various types of springs. Response curves for (b) linear spring, (c) hardening spring, and (d) softening spring. (e) Effect of level of excitation on response.

whence

$$\bar{i} = \frac{p\bar{V}}{p^2L+pR+1/C} \ . \qquad (7.4)$$

An inverse transformation applied to equation (7.4) will give an expression for the current. If we let the p in the Laplace transform be pure imaginary and equal to $\mathrm{i}\omega$ (where ω is now actually the angular frequency) then we obtain the Fourier transform, and thus we can state that C has an 'effective' impedance of $-\mathrm{i}/\omega C$ and L an 'effective' impedance of $\mathrm{i}\omega L$.

Now the Laplace (and Fourier) transforms cannot be used in a nonlinear system. A typical example of a nonlinear system is a unit (for simplicity) inductor in series with a nonlinear resistor whose value is in direct proportion to the current. Let us imagine that a step function of 1 V (at time $t = 0$, voltage rises from 0 to 1 V and remains there) is applied to the system. The equation for the system is:

$$\frac{\mathrm{d}i}{\mathrm{d}t} + i^2 = 1 \ . \qquad (7.5)$$

If we attempt to apply a Laplace transformation, then we do not know the transform for a quantity squared. A Laplace transform also assumes that the principle of superposition applies. The principle of superposition *only* holds for a linear system and is illustrated by figure 7.2. If a linear system for an input $x_{\mathrm{A}}(t)$ gives an output $y_{\mathrm{A}}(t)$ and for an input $x_{\mathrm{B}}(t)$ gives an output $y_{\mathrm{B}}(t)$, then for an input $x_{\mathrm{A}}(t)+x_{\mathrm{B}}(t)$ the output will be $y_{\mathrm{A}}(t)+y_{\mathrm{B}}(t)$. As mentioned above, this principle of superposition does not apply to nonlinear systems. There are fortunately some transforms that can be employed in the solution of nonlinear differential equations, such as the Taylor–Cauchy and Laurent–Cauchy transforms which can provide solutions to nonlinear differential equations by providing a set of recursive equations (the solution of one equation allows one to solve the next and so on). In the Taylor–Cauchy transform time is replaced by a complex variable λ, and the variable written in the λ domain is called $W(\lambda)$. It turns out that the kth derivative (with respect to λ) of $W(\lambda)$,

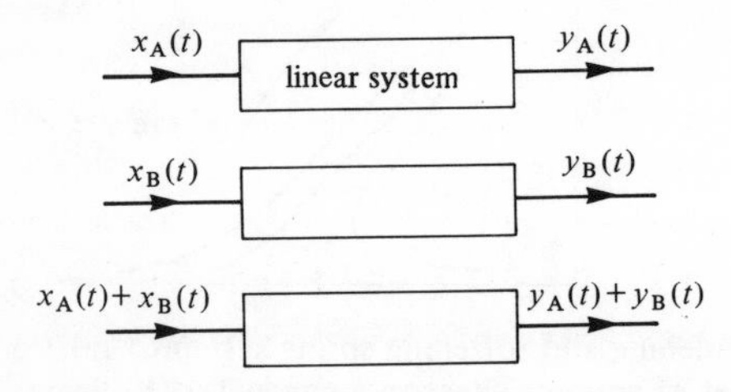

Figure 7.2. Illustration of principle of superposition.

denoted by $W^k(\lambda)$, can be written as a power series in λ:

$$W^k(\lambda) = \sum_{n=0}^{\alpha} w_{n,k}\lambda^n \,, \tag{7.6}$$

where the coefficients $w_{n,k}$ may be found by evaluating the integral

$$w_{n,k} = \frac{1}{2\pi i}\int \frac{W^k(\lambda)}{\lambda^{n+1}}\,d\lambda \,. \tag{7.7}$$

The values of $w_{n,k}$ are found by a series of recursive equations and not by evaluation of the integral in equation (7.7). Expressions can also be found for nonlinear terms such as $[W^k(\lambda)]^n$, $\lambda^n W^k(\lambda)$ etc. in terms of $w_{n,k}$. Our solution thus becomes a solution of a series of equations for $w_{n,k}$. Let us broadly sketch the solution of equation (7.5) using the Taylor–Cauchy approach[2]. Equation (7.5) rewritten in the λ domain is

$$\begin{aligned} W^{(1)}(\lambda)+[W(\lambda)]^2 &= 1 \,, \\ A_0 &= 0 \,. \end{aligned} \tag{7.8}$$

A_0 is a coefficient expressing the initial conditions. Applying the Taylor–Cauchy transform we obtain

$$\begin{aligned} w_n + C^{(2)}_{n-2,0} &= \delta_n \,, \\ C^{(2)}_{n-2,0} &= \sum_{\beta=0}^{n-2} \frac{w_\beta}{\beta+1}\,\frac{w_{n-\beta-2}}{n-\beta-1} \,, \\ W(\lambda) &= \sum_{n=1}^{\infty}\left(\frac{w_{n-1}}{n}\right)^n + A_0 \,. \end{aligned} \tag{7.9}$$

Putting $n = 0$ we get $w_0 = 1$ which then allows us to solve for $n = 1$ and so on, yielding $w_1 = 0$, $w_2 = -1$, $w_3 = 0$, $w_4 = \frac{2}{3}$, $w_5 = 0$, $w_6 = -\frac{17}{45}$, etc. Hence

$$W(\lambda) = \lambda - \tfrac{1}{3}\lambda^3 + \tfrac{2}{15}\lambda^5 - \tfrac{17}{315}\lambda^7 + \dots \,. \tag{7.10}$$

Equation (7.10) happens to be the expansion for $\tanh(\lambda)$, so

$$i(t) = \tanh(t) = \frac{e^t + e^{-t}}{e^t - e^{-t}} \,. \tag{7.11}$$

It is evident that the analysis of a nonlinear system is complicated even when we know the equations governing the system, and certainly the determination of the characteristics of a system from measurements on signals in and out of the system is extremely difficult. If the details of the system are known apart from the exact values of the various

[2] cf R. W. Harris, 1962, "Analysis of pulse in linear and nonlinear systems", *Proc. IRE Aust.*, **23**, 712; Y. H. Ku and A. A. Wolf, 1960, "Taylor-Cauchy transforms for analysis of a class of nonlinear systems", *Proc. IRE*, **48**, 912.

coefficients, then the system output and the output of a simulator (which could be analogue or digital) can be compared, and the coefficients adjusted until the difference is negligible. This approach is pictured in figure 7.3.

For the discussion that follows we shall write the Fourier integral in the form

$$x(t) = \int_{-\infty}^{\infty} e^{i\omega t} dZ(\omega) = \int_{-\infty}^{\infty} e^{i\omega t} X(\omega) d\omega \,, \tag{7.12}$$

where $x(t)$ is the input signal and the integrated power spectrum $dF(\omega)$ [power present over a frequency interval $d(\omega)$] is given by

$$dF(\omega) = \overline{|dZ(\omega)|^2} = \overline{|X(\omega)d\omega|^2} \,. \tag{7.13}$$

The power spectral density is given by:

$$S_{xx}(\omega) = \frac{dF(\omega)}{d\omega} = \overline{|X(\omega)|^2} \,. \tag{7.14}$$

We notice two points in the way the relationship between the time and frequency domain is expressed in equation (7.12). Firstly $dZ(\omega)$ is not a function of time, and secondly the variables contained within the group $\{Z(\omega)\}$ [{ } is a means of referring to the group of numbers made up of $Z(\omega)$ for all values of ω] must be statistically independent, and hence $\{Z(\omega)\}$ is a random process or, alternatively, the members of the group are orthogonal. This orthogonal property is why Fourier analysis works, since for example

$$\int_0^T \sin\omega_1 \sin\omega_2 \, dt = \begin{cases} 0 & (\omega_1 \neq \omega_2) \\ 1 & (\omega_1 = \omega_2) \end{cases} \,. \tag{7.15}$$

If we did not use orthogonal functions then we cannot determine the values of $Z(\omega)$ uniquely.

Now, if $\{Z(\omega)\}$ does not represent a statistically independent group [statistically independent means $Z(\omega_1)$], then equation (7.12) does not hold, and we cannot have a meaningful description of the signal in the frequency domain if $\{Z(\omega)\}$ does not have these attributes. If the system is nonlinear then there exist relationships between the members of the group $\{Z(\omega)\}$, and equation (7.12) is meaningless. Also, if the process is nonstationary, then samples of the signal taken at different times will

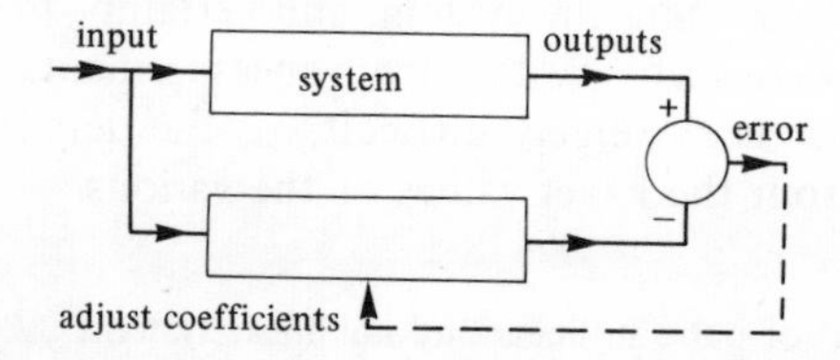

Figure 7.3. Nonlinear system analysis using a simulator.

provide different spectra, and thus, since the $Z(\omega)$ will have to be a function of time, there can exist correlations between the $Z(\omega)$ at different times, and again equation (7.12) does not hold. One could, for a nonstationary process, define a special correlation function $G(\omega, \nu)$:

$$G(\omega, \nu)\,\mathrm{d}\omega\,\mathrm{d}\nu = \overline{\mathrm{d}Z(\omega)\,\mathrm{d}Z^*(\nu)}\,, \tag{7.16}$$

where the bar denotes an average and the asterisk a complex conjugate. It is impossible to imagine what physical significance can be attached to this special correlation function.

If we can subdivide the time interval into a number of sub-intervals, and if $x(t)$ is stationary within each sub-interval, although its statistical properties may vary from time interval to time interval, then we may talk about an 'instantaneous power spectrum' $F_{xx}(\omega, T)$ defined over the subdivision time interval T as

$$F_{xx}(\omega, T) = \left| \int_0^T x(t)\mathrm{e}^{-\mathrm{i}\omega t}\,\mathrm{d}t \right|^2 . \tag{7.17}$$

We must be very careful when specifying this instantaneous power spectrum, because of the uncertainty principle which does not allow us to obtain simultaneously an arbitrarily high degree of resolution in both the time and frequency domain.

A way to overcome the limitations placed by the uncertainty principle, if we try to subdivide the time interval, is to find a new set of functions $\{Z(\omega)\}$ which are now statistically independent (even for a nonstationary system); then we can put

$$x(t) \equiv \int_{-\infty}^{\infty} \phi_t(\omega)\,\mathrm{d}Z(\omega)\,. \tag{7.18}$$

We now have a function $\phi_t(\omega)$ of both time and angular frequency. Mathematically, what is happening in equation (7.18) is that a group of variables is specified which are orthogonal and therefore can be uniquely determined. How unique our choice can be in a nonstationary system is, however, a matter for conjecture. The function $\phi_t(\omega)$ will usually have a Fourier transform, so that we may put

$$\phi_t(\omega) = A_t(\omega)\mathrm{e}^{\mathrm{i}\omega t}\,, \tag{7.19}$$

and hence equation (7.18) becomes

$$x(t) = \int_{-\infty}^{\infty} A_t(\omega)\mathrm{e}^{\mathrm{i}\omega t}\,\mathrm{d}Z(\omega)\,. \tag{7.20}$$

Equation (7.20) defines a meaningful integral for a nonstationary process [provided we can find our set of functions $\{Z(\omega)\}$], and so we can define the 'evolutionary power spectrum' (which is a function of time) as

$$F_{xx}(\omega, t) = |A_t(\omega)|^2|\mathrm{d}Z(\omega)|^2\,. \tag{7.21}$$

If $F_{xx}(\omega, t)$ is differentiable with respect to ω, then we can define the 'evolutionary spectral density function':

$$S_{xx}(\omega, t) = \frac{\mathrm{d}F_{xx}(\omega, t)}{\mathrm{d}\omega} . \tag{7.22}$$

Unfortunately, the state of the art of evolutionary spectra is not completely developed to provide a way of determining the group $\{Z(\omega)\}$. One suggested way of determining an evolutionary spectrum is to pass the signal through a filter centred on some frequency, say, ω_0, and having a frequency response function $H(\omega)$ normalised so that

$$\int_{-\infty}^{\infty} |H(\omega)|^2 \mathrm{d}\omega = 1 . \tag{7.23}$$

Let the output of the filter be denoted by $Y(t)$; the values of $Y^2(t)$ are then smoothed by a weight function $W_{T'}(u)$ chosen so that

$$\lim_{T' \to \infty} T' \int_{-\infty}^{\infty} |W_{T'}^{F}(\omega)|^2 \mathrm{d}\omega = C , \tag{7.24}$$

where $W_{T'}^{F}(\omega)$ is the Fourier transform of $W_{T'}(u)$ and C is a constant. Then the estimate of the evolutionary spectrum is

$$\hat{F}_{xx}(\omega_0, t) = \int_{-\infty}^{\infty} W_{T'}(u) | Y^2(t-u) | \mathrm{d}u , \tag{7.25}$$

where ^ denotes an estimate. A suggested choice of weight function is

$$W_{T'}(u) = \begin{cases} \dfrac{1}{T'} & -\tfrac{1}{2}T' < u < \tfrac{1}{2}T' \\ 0 & \text{otherwise} , \end{cases} \tag{7.26}$$

whence $C = 2\pi$.

7.3 Nonlinear systems and the p.d.f.

Spectral analysis can yield little information about a nonlinear system as the values of the cross-spectral density or output spectral density do not form statistically independent sets. For example, if the output was related to the input by $y = ax + bx^2$, then a simple calculation shows that the amplitude at the output, at frequency 2ω, depends on both the frequencies ω and 2ω at the input. We find some interesting results if we examine the p.d.f. at the input and output for a nonlinear system. If $x(t)$ is the input signal and $y(t)$ the output signal, then we know the total probability of finding any signal amplitude is 1, whence

$$\int_{-\infty}^{\infty} p(x) \mathrm{d}x = \int_{-\infty}^{\infty} p(y) \mathrm{d}y . \tag{7.27}$$

Differentiating both sides with respect to $\mathrm{d}x\,\mathrm{d}y$ and rearranging we obtain

$$p(y) = p(x) \Big/ \left| \frac{\mathrm{d}y}{\mathrm{d}x} \right| . \tag{7.28}$$

The modulus sign around the derivative is required to ensure that $p(y)$ cannot become negative. Note that in the derivation of equation (7.27) we did not need to assume that the system was linear. Thus we have a relationship between the p.d.f. of the input and the output and, if the system is linear, the p.d.f's are the same (except, possibly, for a constant factor). Thus for $y = cx$ (linear system) we have

$$p(y) = \frac{p(x)}{|c|} . \tag{7.29}$$

We now have a means of determining whether a system is nonlinear or not by comparing the p.d.f's, and we can determine $|\mathrm{d}y/\mathrm{d}x|$ from their ratio. It is conceivable that the characteristic function may provide more insight into the situation, but the behaviour of the characteristic function in a nonlinear situation has as yet to be determined. To give some idea as to what happens let us look at the p.d.f. of a signal having a Gaussian profile

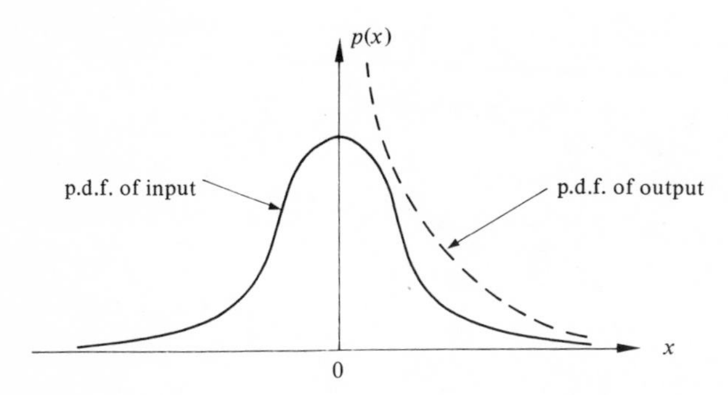

Figure 7.4. P.d.f. for input and output of a square-law device.

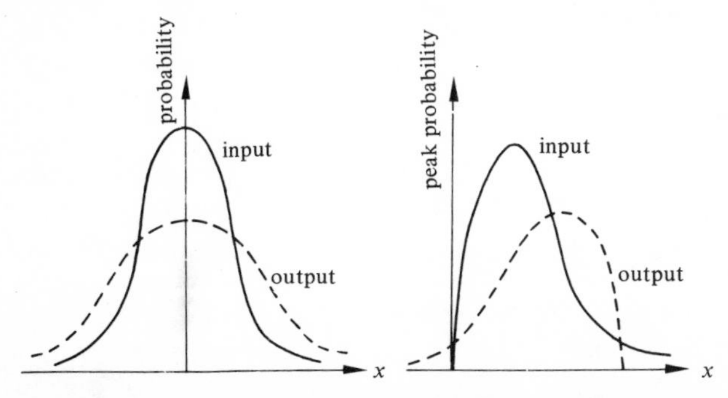

Figure 7.5. Comparison of input and output probability distributions for a hardening spring system.

which is passed through a square law device of the form $y = ax^2$, as shown in figure 7.4.

Another concept which has been employed, but is more difficult to interpret, is the peak probability distribution function, where we examine the probability distribution of the signal peaks (or maxima). To illustrate this, figure 7.5 shows the p.d.f. and the peak p.d.f. for the input and output of a system having a hardening spring where the input signal has a Gaussian characteristic.

The peak p.d.f. for a linear system having a Gaussian probability distribution function is a Rayleigh distribution (see section 1.2.3).

8

The statistics of functions of random variables

8.1 General
In this chapter we shall consider the usual statistical functions of a function $z(t)$ which may be regarded as a sum or a product of two random variables $x(t)$ and $y(t)$.

8.2 Autocorrelation of the sum of two random variables
If

$$z(t) = x(t)+y(t)\,, \tag{8.1}$$

then, applying the usual definition for the autocorrelation of $z(t)$, we have

$$R_{zz}(\tau) = \lim_{T\to\infty} \frac{1}{T}\int_0^T z(t)z(t+\tau)\,\mathrm{d}t\,. \tag{8.2}$$

Substituting for $z(t)$ from (8.1) into (8.2) we obtain

$$R_{zz}(\tau) = \lim_{T\to\infty} \frac{1}{T}\int_0^T [x(t)+y(t)][x(t+\tau)+y(t+\tau)]\,\mathrm{d}t\,, \tag{8.3}$$

which can be written as four separate integrals as follows:

$$R_{zz}(\tau) = \lim_{T\to\infty} \frac{1}{T}\int_0^T x(t)x(t+\tau)\,\mathrm{d}t + \lim_{T\to\infty} \frac{1}{T}\int_0^T x(t)y(t+\tau)\,\mathrm{d}t$$
$$+\lim_{T\to\infty} \frac{1}{T}\int_0^T y(t)x(t+\tau)\,\mathrm{d}t + \lim_{T\to\infty} \frac{1}{T}\int_0^T y(t)y(t+\tau)\,\mathrm{d}t\,. \tag{8.4}$$

Each of the four integrals may now be recognised as a correlation function and hence we may write

$$R_{zz}(\tau) = R_{xx}(\tau)+R_{xy}(\tau)+R_{yx}(\tau)+R_{yy}(\tau)\,. \tag{8.5}$$

Remember that $R_{xy}(\tau)$ and $R_{yx}(\tau)$ are in general different and expression (8.5) may only be simplified if $x(t)$ and $y(t)$ are uncorrelated, in which case $R_{zz}(\tau)$ is the sum of the two separate autocorrelation functions.

8.3 Power spectral density of the sum of two random variables
Using the result of section 8.2 and applying the Wiener–Khintchine relationships between power spectral density and correlation functions we obtain immediately

$$S_{zz}(\omega) = S_{xx}(\omega)+S_{xy}(\omega)+S_{yx}(\omega)+S_{yy}(\omega)\,. \tag{8.6}$$

It is however of interest to obtain this result in an independent manner as follows. If $z(t) = x(t)+y(t)$, then

$$z_T(t) = x_T(t)+y_T(t)\,, \tag{8.7}$$

where the suffix T implies a sample of length T has been taken from the original sample.

Taking Fourier transforms of each side of equation (8.7) we have

$$Z_T(\mathrm{i}\omega) = X_T(\mathrm{i}\omega) + Y_T(\mathrm{i}\omega) \; ; \tag{8.8}$$

simple complex number theory gives the complex conjugate as

$$Z_T^*(\mathrm{i}\omega) = X_T^*(\mathrm{i}\omega) + Y_T^*(\mathrm{i}\omega) \, . \tag{8.9}$$

Applying the usual definition for the power spectral density of $z(t)$ we have

$$S_{zz}(\omega) = \lim_{T\to\infty} \frac{1}{T} Z_T(\mathrm{i}\omega) Z_T^*(\mathrm{i}\omega) \, . \tag{8.10}$$

Substituting for $Z_T(\mathrm{i}\omega)$ and $Z_T^*(\mathrm{i}\omega)$ we have

$$S_{zz}(\omega) = \lim_{T\to\infty} \frac{1}{T} [X_T(\mathrm{i}\omega) + Y_T(\mathrm{i}\omega)][X_T^*(\mathrm{i}\omega) + Y_T^*(\mathrm{i}\omega)] \, . \tag{8.11}$$

We may now expand the right hand side of equation (8.11) into four separate terms as follows:

$$S_{zz}(\omega) = \lim_{T\to\infty} \frac{1}{T} X_T(\mathrm{i}\omega) X_T^*(\mathrm{i}\omega) + \lim_{T\to\infty} \frac{1}{T} X_T(\mathrm{i}\omega) Y_T^*(\mathrm{i}\omega)$$
$$+ \lim_{T\to\infty} \frac{1}{T} Y_T(\mathrm{i}\omega) X_T^*(\mathrm{i}\omega) + \lim_{T\to\infty} \frac{1}{T} Y_T(\mathrm{i}\omega) Y_T^*(\mathrm{i}\omega) \, . \tag{8.12}$$

This can now be recognised as containing four separate spectral functions and hence is equivalent to equation (8.6).

8.4 Probability density function of the sum of two independent random variables

If $z(t) = x(t) + y(t)$, we can write

$$\mathrm{f}(x,y) = \mathrm{f}_1(x)\mathrm{f}_2(y) = \mathrm{f}(x,y) = \mathrm{f}(x)\mathrm{f}_2(z-x) \, . \tag{8.13}$$

Now the probability that $z \leqslant Z$ is

$$\mathrm{F}(Z) = \mathrm{F}(y \leqslant Z - x) \, , \tag{8.14}$$

i.e.

$$\mathrm{F}(z \leqslant Z) = \int_{-\infty}^{\infty} \int_{-\infty}^{Z-x} \mathrm{f}(x,y)\, \mathrm{d}x\, \mathrm{d}y \, , \tag{8.15}$$

which on differentiating with respect to Z yields

$$\mathrm{f}(z) = \int_{-\infty}^{\infty} \mathrm{f}_1(x)\mathrm{f}_2(y)\, \mathrm{d}x \, . \tag{8.16}$$

As $y = Z - x$, this becomes

$$f(z) = \int_{-\infty}^{\infty} f_1(x) f_2(Z-x)\,dx\,. \tag{8.17}$$

This expression may be recognised as the convolution integral so that we may write

$$f(z) = f_1(x) * f_2(x)\,. \tag{8.18}$$

8.5 Autocorrelation of the product of two random signals

As before we take the new random variable

$$z(t) = x(t)y(t)\,. \tag{8.19}$$

This situation occurs in many practical problems, for example, the modification of one signal by another.

Proceeding with the usual definition for the autocorrelation function $R_{zz}(\lambda)$ we have

$$R_{zz}(\lambda) = \lim_{T\to\infty} \frac{1}{T}\int_0^T x(t)y(t)x(t+\lambda)y(t+\lambda)\,dt\,. \tag{8.20}$$

Multiplying each side of (8.20) by itself and omitting the limit signs for clarity we have

$$R_{xx}^2(\lambda) = \left[\int_0^T x(t)y(t)x(t+\lambda)y(t+\lambda)\,dt\right]^2. \tag{8.21}$$

Assuming that the product of the two integrals may be written as a double integral, we obtain after some rearranging

$$R_{zz}^2(\lambda) = \iint x^2(t)x^2(t+\lambda)y^2(t)y^2(t+\lambda)\,dt\,dt\,, \tag{8.22}$$

which can be written

$$R_{zz}^2(\lambda) = \int x^2(t)x^2(t+\lambda)\,dt \int y^2(t)y^2(t+\lambda)\,dt\,. \tag{8.23}$$

The two integrals may be recognised as the autocorrelation of the square of the signal, whence it follows that

$$R_{zz}^2(\lambda) = R_{x^2x^2}(\lambda)R_{y^2y^2}(\lambda)\,. \tag{8.24}$$

As a simple example consider the case of a sine wave modulated by a signal which switches from 0 to 1 at regular intervals. In this case

$$x(t) = \sin\omega t\,, \tag{8.25}$$

and

$$x^2(t) = \tfrac{1}{2}(1+\cos 2\omega t)\,, \tag{8.26}$$

$$R_{x^2x^2}(\lambda) = \tfrac{1}{2}(1+\cos 2\lambda)\,. \tag{8.27}$$

This situation is shown in figure 8.1 and the resulting correlation functions are shown in figure 8.2.

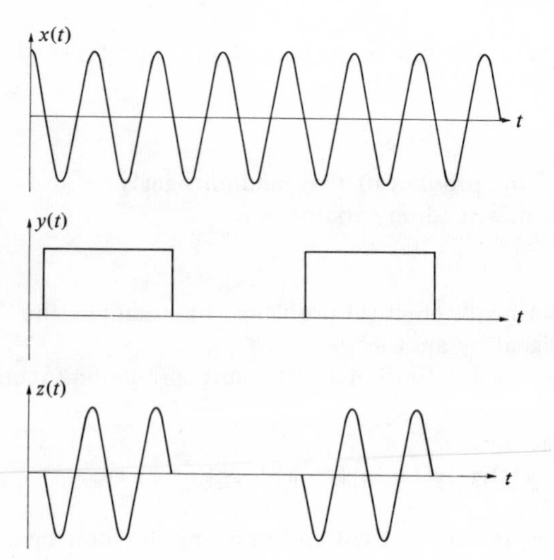

Figure 8.1. Modulation of a sine wave by a gating signal.

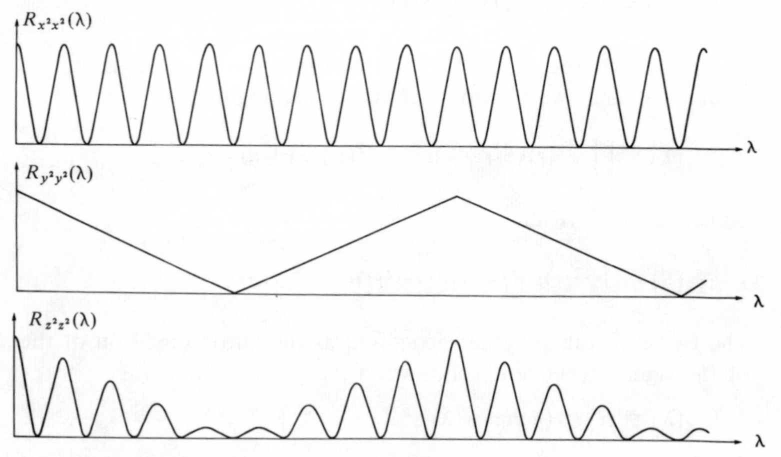

Figure 8.2. The square of the autocorrelation function of the product of a sine wave and a gating function.

Appendix I

Glossary of terms

Aliasing. Occurs in digitally sampled data when an input signal is sampled at intervals of $\Delta\tau$. The Nyquist frequency is $1/2\Delta\tau$ and frequencies which are a multiple of the Nyquist frequency cannot be distinguished from it. This effect is called aliasing.

Average. If samples $x_1, x_2, x_3, ..., x_n$ exist with corresponding frequencies of occurrence $f_1, f_2, f_3, ..., f_n$, then the average is defined as

$$\bar{x} = \frac{\sum f_i x_i}{n} .$$

For an analogue signal $x(t)$, the *time* average is defined as

$$\bar{x} = \frac{1}{T}\int_0^T x(t)\,dt .$$

If an ensemble of records $x_1(t), x_2(t), ..., x_n(t)$ are sampled at a specific time τ then the *ensemble* average is

$$\tilde{x} = \frac{\sum x_i(\tau)}{n} .$$

Band-limited noise. A term usually taken to mean that a broad-band signal has been passed through a band-pass filter. Quite often used when the lower frequency of the band-pass filter is 0 Hz.

Bandwidth. Usually taken, for a real filter, as the width of the frequency response curve of the filter at the 3 dB points. Sometimes this width is expressed as a fraction of the centre frequency of the filter.

Bivariate distribution. If two random variables y_1 and y_2, having variances of σ_1 and σ_2 with a covariance σ_{12}, have a joint probability distribution function of the form

$$p(y_1, y_2) = \frac{1}{2\pi(\sigma_1\sigma_2 - \sigma_{12})}\exp\left[-\frac{\sigma_1 y_1^2 + 2\sigma_{12} y_1 y_2 - \sigma_2 y_2^2}{2(\sigma_1\sigma_2 - \sigma_{12})}\right],$$

then they are said to have a bivariate Gaussian distribution.

Binomial distribution. Discovered in 1700 by James Bernoulli, this basic distribution arises in the consideration of the relative frequency of events which is governed by the expression $(p+q)^n$, where p is the probability of success of the event.

Characteristic function $\phi(\alpha)$. Similar in form to Laplace or Fourier transform in that the characteristic function $\phi(\alpha)$ is an infinite integral transform of the probability density function $f(x)$ with $e^{-i\alpha x}$ as the kernel:

$$\phi(\alpha) = \int_{-\infty}^{\infty} f(x) e^{-i\alpha x}\,dx .$$

Coherence function $C_{xy}(\omega)$. Measures the degree of linear dependence between a pair of signals $x(t)$ and $y(t)$ having power spectral densities $S_{xx}(\omega)$ and $S_{yy}(\omega)$, respectively, and cross spectral density $S_{xy}(\omega)$:

$$C_{xy}(\omega) = \frac{|S_{xy}(\omega)|^2}{S_{xx}(\omega)S_{yy}(\omega)} \leqslant 1 .$$

Conditional probability $P(\mathrm{B}/\mathrm{A})$. The probability of the occurrence of event B given that event A has occurred is defined as the ratio of the joint probability $P(\mathrm{A}, \mathrm{B})$ to the probability of the occurrence of event A:

$$P(\mathrm{B}/\mathrm{A}) = \frac{P(\mathrm{A}, \mathrm{B})}{P(\mathrm{A})} .$$

Convolution. The process in which one function, say $a(\lambda)$, is convolved or folded back along another function $b(t)$. If the result of the convolution is $c(t)$ we define the process as follows:

$$c(t) = \int_{-\infty}^{\infty} a(\lambda)b(t-\lambda)\,\mathrm{d}\lambda ,$$

sometimes written $c = a^* b$.

Correlation, auto, $R_{xx}(\lambda)$. The average value of a function $x(t)$ multiplied by a delayed version of itself. This average value is a function of the amount of delay λ introduced:

$$R_{xx}(\lambda) = \lim_{T\to\infty} \frac{1}{2T}\int_{-T}^{T} x(t)x(t-\lambda)\,\mathrm{d}t .$$

Correlation, cross, $R_{xy}(\lambda)$. The average value of the product of two functions $x(t)$ and $y(t-\lambda)$ where λ is an adjustable delay introduced between the two signals:

$$R_{xy}(\lambda) = \lim_{T\to\infty} \frac{1}{2T}\int_{-T}^{T} x(t)y(t-\lambda)\,\mathrm{d}t .$$

Correlation coefficient $\sigma_{xy}(\lambda)$. A measure of the degree of correlation between two signals $x(t)$ and $y(t)$ having autocorrelation functions $R_{xx}(\lambda)$ and $R_{yy}(\lambda)$ together with a cross correlation coefficient $R_{xy}(\lambda)$:

$$\sigma_{xy}(\lambda) = \frac{R_{xy}(\lambda)}{[R_{xx}(0)R_{yy}(0)]^{1/2}} \qquad 0 \leqslant |\sigma_{xy}(\lambda)| \leqslant 1 .$$

Covariance. The covariance of random signals x and y is the statistical average (or expectation) of the product of the fluctuations of the variables about their respective means.

$$\text{Covariance} = E[(x-\bar{x})(y-\bar{y})]$$

$$= \int_{-\infty}^{\infty}\int_{-\infty}^{\infty}(x-\bar{x})(y-\bar{y})p(x,y)\,\mathrm{d}x\,\mathrm{d}y$$

$$= \lim_{T\to\infty}\frac{1}{2T}\int_{-T}^{T}(x-\bar{x})(y-\bar{y})\,\mathrm{d}t\,.$$

Cross-spectral density $S_{xy}(\omega)$. Measure of the relationship between a pair of signals in the frequency domain. The cross-spectral density is also the Fourier transform of the cross-correlation function. If $X_T(\mathrm{i}\omega)$ and $Y_T(\mathrm{i}\omega)$ are the Fourier transforms of signals $x(t)$ and $y(t)$ of sample length T, then cross-spectral density is defined as

$$S_{xy}(\omega) = \lim_{T\to\infty}\frac{1}{2T}X_T(\mathrm{i}\omega)Y_T^*(\mathrm{i}\omega)\,;$$

also

$$S_{xy}(\omega) = \int_{-\infty}^{\infty}R_{xy}(\lambda)\mathrm{e}^{-\mathrm{i}\omega\lambda}\,\mathrm{d}\lambda\,.$$

Delta function $\delta(t)$. Sometimes called the impulse function; its value is zero everywhere except at the origin where it is infinite. The area under the delta function is unity.

$$\int_{-\infty}^{\infty}\delta(t)\,\mathrm{d}t \equiv 1\,.$$

Deterministic. A term applied to signals or functions whose value at some future time can be predicted precisely from a knowledge of past events; for example sine waves and square waves are deterministic signals.

Discrete random variable. A random variable x is a discrete random variable if it has only a finite number of values in any finite time interval. For example the number of heads in a coin tossing experiment is a discrete random variable.

Ensemble average. If a collection of records $x_1(t), x_2(t), x_3(t), \ldots, x_n(t)$, are sampled at the same time, say t_1, then the ensemble average is the mean of the sampled records:

$$\widetilde{x(t_1)} = \frac{1}{n}\sum_n x_i(t_1)\,.$$

Ergodic. If the time average of a random signal

$$\bar{x} = \frac{1}{T}\int_{-\infty}^{\infty}x(t)\,\mathrm{d}t$$

is equal to the ensemble average then the process is called ergodic.

Evolutionary spectra. When the spectrum associated with a random variable $x(t)$ is a function of both ω the angular frequency and a parameter p, i.e. P.S.D. $= S_{xx}(\omega, p)$, so that $S_{xx}(\omega) = f(p)$, then the spectrum is said to be evolutionary.

Fourier transform. Similar to Laplace transform and is defined by the integral transform

$$X(\mathrm{i}\omega) = \int_{-\infty}^{\infty} x(t)\mathrm{e}^{-\mathrm{i}\omega t}\mathrm{d}t\ ,$$

where $X(\mathrm{i}\omega)$ is the Fourier transform of $x(t)$.

Frequency response $H(\omega)$. If a linear system is excited by a sine wave input then the output is also a sine wave, in general of different amplitude and phase with respect to the input. A plot of this amplitude and phase information constitutes the frequency response of the system.

Gaussian distribution. The familiar bell-shaped distribution whose probability density function is given by

$$p(x) = \frac{1}{(2\pi\sigma)^{\frac{1}{2}}}\exp\left[-\frac{(x-\overline{x})^2}{2\sigma^2}\right],$$

where $\overline{x}$ is the mean and σ^2 the variance.

Impulse response $h(t)$. The response of a system to a delta function input:

$$H(\omega) = \int_{-\infty}^{\infty} h(t)\mathrm{e}^{-\mathrm{i}t}\mathrm{d}t\ .$$

The impulse response and the frequency response form a Fourier transform pair.

Kernel. In an integral transform defined as

$$F(p) = \int_{a}^{b} f(t)K(t,p)\mathrm{d}t$$

$K(t,p)$ is the kernel of the transform and can take many different forms. For example, the kernel of a Laplace transform is e^{-tp}.

Laurent–Cauchy transform. An integral transform which can be employed to find solutions to nonlinear differential equations. The transformed variables are written as power series in time. Thus, if $H(t)$ is a function of time,

$$H(t) = \sum_{n=0}^{\infty} h_n t^{-n}\ ,$$

where

$$h_n = \frac{1}{2\pi j}\int H(t)t^{n-1}\mathrm{d}t\ .$$

Linear system. A system is linear when the principle of linear superposition holds: if the output is $y_1(t)$ for an input $x_1(t)$ and $y_2(t)$ for an input $x_2(t)$, then an input $x_1(t)+x_2(t)$ would for a linear system yield $y_1(t)+y_2(t)$ as an output.

Markovian process. A Markovian (Markoff) process is one in which the future state of the process depends only on the present state and not on the route at which the present state was arrived at.
A Markovian process can be shown to have an autocorrelation function of the form of a decaying exponential function:

$$R_{xx}(\tau) = R_{xx}(0)\mathrm{e}^{-|\tau|/\lambda}.$$

Mean and mean square. Mean is defined as

$$\bar{x} = \frac{1}{T}\int_0^T x(t)\,\mathrm{d}t\,,$$

mean square as

$$\overline{x^2} = \frac{1}{T}\int_0^T x^2(t)\mathrm{d}t\,.$$

Moment generating function $\phi(\alpha)$. The moment generating function is the same as the characteristic function. The rth moment of a distribution is given by

$$\mu_r = \frac{1}{(-\mathrm{i})^r}\left[\frac{\mathrm{d}^r\phi(\alpha)}{\mathrm{d}\alpha^r}\right]_{\alpha=0}.$$

Narrow band. If a signal has the characteristic of one in which one particular frequency dominates, then the process is called narrow band. A process is narrow band if the width Δf of the significant part of the power spectral density is small compared to the centre frequency. A narrow band process may be denoted as

$$x(t) = y(t)\cos[W_0 t+\phi(t)]\,.$$

Nonlinear. A system is said to be nonlinear if the superposition principle does not apply. For example, a stimulus $x(t)$ produces a response $y(t)$; then for a nonlinear system a stimulus $2x(t)$ would not necessarily produce a response $2y(t)$.

Nyquist frequency. This is the highest frequency which can be determined in a Fourier analysis of a discrete sampling of data. If a time series is sampled at interval Δt, this frequency is usually taken to be $1/(2\Delta t)$ Hz, or more conservatively, $1/(4\Delta t)$ Hz.

Parseval's theorem. This relates the mean square of a signal with the integral of the power spectral density

$$\lim_{T\to\infty}\frac{1}{2T}\int_{-T}^{T} x^2(t)\mathrm{d}t = \frac{1}{2\pi}\int_{-\infty}^{\infty} S_{xx}(\omega)\mathrm{d}\omega\,.$$

Poisson distribution. The limit of the binomial distribution when the probability p is small.

$$\left.\begin{array}{l} p \rightarrow 0 \\ np \rightarrow m, \text{ say} \end{array}\right\} \text{ as } n \rightarrow \infty ,$$

$$P(r) = \frac{m^r}{r!} \mathrm{e}^{-m} .$$

The mean and variance of a Poisson distribution are both equal to m.

Power spectral density (P.S.D.) $S_{xx}(\omega)$. Measures how a signal is distributed in the frequency domain:

$$S_{xx}(\omega) = \lim_{T \rightarrow \infty} \frac{1}{2T} X_T(\mathrm{i}\omega) X_T^*(\mathrm{i}\omega) .$$

Probability density function (p.d.f.) $\mathrm{f}(x)$. The probability that a signal $x(t)$ lies in the range $X \pm \frac{1}{2}\delta X$ is simply $\mathrm{f}(x)\delta x$. For an ergodic process the fraction of time a signal spends in the δX wide 'window' is $\mathrm{f}(x)\delta x$.

Probability, cumulative. $\mathrm{F}(x)$ is the integral of the probability density function:

$$\mathrm{F}(x) = \int_{-\infty}^{x} \mathrm{f}(x)\,\mathrm{d}x .$$

The probability that a signal lies below a level A is simply $\mathrm{F}(A)$.

Probability, joint. $\mathrm{F}(X_1, X_2)$ and $\mathrm{f}(x_1, x_2)$. If X_1 and X_2 are two random signals then $\mathrm{F}(X_1, X_2)$ is the probability that both $x_1 \leqslant X_1$ and $x_2 \leqslant X_2$;

$$\mathrm{f}(X_1, X_2) = \frac{\partial \mathrm{F}(X_1, X_2)}{\partial X_1 \partial X_2} .$$

Rayleigh distribution. A probability density function named after Lord Rayleigh.

$$\mathrm{f}(x) = \frac{2x}{\sigma^2} \exp\left(-\frac{x^2}{\sigma^2}\right) \text{ for } x > 0$$

$$\mathrm{f}(x) = 0 \quad \text{otherwise.}$$

If a Gaussian process is passed through a narrow band filter, then the p.d.f. of the envelope is a Rayleigh distribution.

Recursive. This term when applied to digital filters means that the current output depends on the current input and a linear combination of past inputs *and* outputs. A non-recursive filter output depends on the current input and a linear combination of past inputs only.

Root mean square (r.m.s.).

$$\text{r.m.s.} = \left[\frac{1}{T}\int_0^T x^2(t)\,\mathrm{d}t\right]^{1/2} .$$

Sinc function.

$$\operatorname{sinc}X = \frac{\sin X}{X} .$$

Signal to noise ratio. Usually taken as the ratio between the r.m.s. of the signal and the r.m.s. of unwanted noise.

Standard deviation σ. This is the square root of the variance σ^2 and is defined as

$$\sigma = \left[\frac{(x-\bar{x})^2}{n-1}\right]^{1/2} .$$

Statistical independence. For statistically independent events the joint probability density function is equal to the product of the individual probability functions,

$$\mathrm{f}(x_1, x_2) = \mathrm{f}(x_1)\mathrm{f}(x_2) ,$$

if and only if x_1 and x_2 are statistically independent.

Stationary. A signal is said to be stationary when all its statistical descriptions are invariant with time.

Stochastic. A stochastic process is one in which the phenomena are governed by probabilistic laws.

Taylor–Cauchy transform. This allows a function of time to be expressed as a power series. Letting λ replace time, we define the transformed coefficients w_n of a function $W(\lambda)$ as

$$w_n = \frac{1}{2\pi j}\int \frac{W(\lambda)}{\lambda^{n+1}}\mathrm{d}\lambda ,$$

where

$$W(\lambda) = \sum_{n=0}^{\infty} w_n \lambda^n .$$

Transform function $H(\omega)$. Relates the input and output of a linear system, and also is the Fourier transform of the impulse response $h(t)$:

$$H(\omega) = \int_{-\infty}^{\infty} h(t)\mathrm{e}^{-\mathrm{i}\omega t}\,\mathrm{d}t .$$

If the input spectrum is $S_{xx}(\omega)$ and the cross spectrum is $S_{xy}(\omega)$, then

$$H(\omega) = \frac{S_{xy}(\omega)}{S_{xx}(\omega)} .$$

Truncated. If a signal $x(t)$ is set to zero outside a given range $\pm T$ and retains its original value within that range, it is said to be truncated.

Variance σ^2. This is the square of the standard deviation σ

$$\sigma^2 = E(x-\overline{x})^2 ,$$

$$= \frac{1}{T}\int_0^T (x-\overline{x})^2 \mathrm{d}t .$$

Wiener–Khintchine relations. These relate the power spectral density and correlation functions

$$S_{xy}(\omega) = \int_{-\infty}^{\infty} R_{xy}(\tau)\mathrm{e}^{-\mathrm{i}\omega t}\,\mathrm{d}t$$

$$R_{xy}(\tau) = \frac{1}{2\pi}\int_{-\infty}^{\infty} S_{xy}(\omega)\mathrm{e}^{-\mathrm{i}\omega\tau}\,\mathrm{d}\omega .$$

White noise. A concept in which the power spectral density is a constant over all the frequency range. This clearly implies infinite power which is physically unrealisable. However, some systems can be 'white' over part of the frequency range of interest.

Appendix II

Introduction to binary systems

The number system that we normally use is the decimal system where each digit can take one of ten values 0, 1, 2, ..., 9. If we were to use the decimal system in a computer, then each digit would be represented by a voltage, and this voltage would have to take one of a possible ten different values. Should some minor malfunction occur in the electronics, then it is possible for, say, an 8 to be misrepresented as a 7 or a 9. A number system based on 10 is not a convenient system for electronic computers, and a better system uses a base of 2, and is known as the binary system. The only digits required now are 0 and 1, and a slight malfunction could now not so easily lead to an error as one only has to determine whether a voltage is there (1) or not there (0). The processes of addition and multiplication are greatly simplified in a binary system. The binary digits are referred to as bits.

To show how a binary number relates to a decimal number, some examples are given below:

binary number	1	0	0	1	0	1		
decimal equivalent	2^5+			2^2+		2^0		= 37
binary number	1	1	0	0	1	0	0	
decimal equivalent	2^6+	2^5+			2^2			= 100

Addition is simply $1+0 = 1$ and $1+1 = 0$ and carry 1, as exemplified by the example given below:

```
     1 1 0 0 1 1    51
+    0 1 1 0 1 0  + 26
    ←1 ←1     ←1
   1 0 0 1 1 0 1    77
```

To multiply add, shift and add if 1 present, if not shift again, and so on as illustrated by the example given below:

```
         1 1 0 1 0 × 1 1 0 1   26 × 13
         1 1 0 1 0          1:  add, then shift
               0          0:  shift again
     1 1 0 1 0          1:  add, then shift
   1 1 0 1 0          1:  add
 ← ← ← ← ←
 1 0 1 0 1 0 0 1 0                       = 338
```

It is not the intention to proceed any further with the discussion on binary arithmetic, as it has been illustrated how this number system can be most readily adapted for computer operation because of its simplicity and minimal chance of error, since we only have to deal with a voltage which is either there or not there. An A/D converter then has the task of converting signals to a binary form so that they can be handled by a computer. There are two ways of converting signals to a binary system: pure binary and binary coded decimal (BCD). In the BCD system each decimal digit, rather than the entire number, is given in binary form. This is illustrated by the example given below:

decimal	3	5	6
BCD	|0011|	0101|	0110|
pure binary	10 11 00 1 00		

We see that 9 digits are required in pure binary while 3 groups of 4 are required in BCD. However, it is often easier to transmit a sequence of groups of 4 (for example, using punched paper tape) than the entire binary number.

Appendix III

An introduction to the fast Fourier transform (F.F.T.)

It is not the intention of this appendix to provide complete details of the various versions of the F.F.T. but rather to show how one version of this algorithm operates. As mentioned in chapter 3, its great advantages lie in the large reduction in the number of operations required to produce a complete Fourier analysis.

The discrete Fourier transform (D.F.T.) is a Fourier transform using discrete data, and can be defined as

$$C_r = \sum_{k=0}^{N-1} X_k \mathrm{e}^{-2\pi \mathrm{i} rk/N} , \qquad r = 0, 1, 2, ..., N-1 , \tag{III.1}$$

where X_k are the N discrete data points equidistant in time, $\mathrm{i} = (-1)^{\frac{1}{2}}$, and C_r is the Fourier transform, which is a complex number. There is not universal agreement as to the factor to be used in front of the summation sign on the right hand side of equation (III.1), and we have used unity for convenience; however, often the factor $1/N$ is employed and sometimes $(1/N)^{\frac{1}{2}}$. The coefficients of C_r may be written in terms of their real and imaginary parts:

$$C_r = A_r + \mathrm{i}B_r , \tag{III.2}$$

where A_r can be identified with the cosine coefficients and B_r with the sine coefficients. For notational convenience we can introduce a new symbol W such that

$$C_r = \sum_{k=0}^{N-1} X_k W^{rk} , \quad r = 0, 1, ..., N-1 \tag{III.3}$$

where

$$W = \mathrm{e}^{-2\pi \mathrm{i}/N} .$$

The calculation of C_r requires N^2 operations; careful consideration shows, however, that there is a lot of redundancy in the calculations. A great deal of simplification results if the signal is decimated in time as proposed by Cooley and Tukey[3]. We suppose that the time function X_k of N samples is divided into two series of $\frac{1}{2}N$ samples, Y_k and Z_k, such that Y_k is composed of the even-numbered points and Z_k of the odd-numbered points.

$$\left.\begin{aligned} Y_k &= X_{2k} \\ Z_k &= X_{2k+1} \end{aligned}\right\} \quad k = 0, 1, 2, ..., \tfrac{1}{2}N - 1 . \tag{III.4}$$

[3] J. W. Cooley and J. W. Tukey, 1968, "An algorithm for the machine calculation of complex Fourier series", *Math. of Comput.*, **119**, 297–301; see also special issue of IEEE Transactions on Audio and Electroacoustics, June 1967.

Each of these two new sequences will have its own Fourier transform defined by

$$C_r^Y = \sum_{k=0}^{\frac{1}{2}N-1} Y_k \, e^{-4irk/N}$$

$$C_r^Z = \sum_{k=0}^{\frac{1}{2}N-1} Z_k \, e^{-4irk/N} \,. \tag{III.5}$$

Now, we require the D.F.T., i.e. C_r, of the original series X_k, so let us see how this transform is related to the transforms C_r^Y, C_r^Z of the two new series.

$$C_r = \sum_{k=0}^{\frac{1}{2}N-1} [Y_k \, e^{(-2\pi ir/N)2k} + Z_k \, e^{(-2\pi ir/N)(2k+1)}] \,, \tag{III.6}$$

$$= \sum_{k=0}^{\frac{1}{2}N-1} Y_k \, e^{-4\pi irk/N} + e^{-2\pi ir/N} \sum_{k=0}^{\frac{1}{2}N-1} Z_k \, e^{-4\pi irk/N} \,. \tag{III.7}$$

Using equation (III.5) we obtain

$$C_r = C_r^Y + e^{-2\pi ir/N} C_r^Z \,. \tag{III.8}$$

For $r > \frac{1}{2}N$ the Fourier coefficients C_r^Y and C_r^Z must simply repeat periodically. Thus, substituting $r + \frac{1}{2}N$ for r, we find

$$C_{r+\frac{1}{2}N} = C_r^Y + e^{-2\pi i(r+\frac{1}{2}N)/N} C_r^Z = C_r^Y - e^{-2\pi ir/N} C_r^Z \,, \tag{III.9}$$

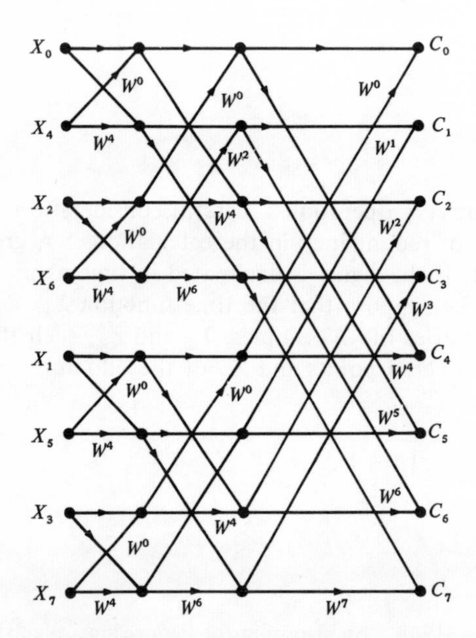

Figure III.1. Signal flow graph illustrating the computation of the D.F.T. when the operations involved are completely reduced to multiplications and additions.

since $e^{-\pi i} = -1$. Using the notation for W we can write

$$C_r = C_r^Y + W^r C_r^Z \; ; \quad C_{r+\frac{1}{2}N} = C_r^Y - W^r C_r^Z \, . \tag{III.10}$$

Thus, with the aid of equations (III.10) the D.F.T. of a sequence X_k of N points can be found from the D.F.T. of two sequences, Y_k and Z_k, each consisting of $\frac{1}{2}N$ points. Each of the sequences Y_k and Z_k can in turn be broken down into two new sequences each of $\frac{1}{4}N$ points and equations like (III.10) employed; these new sequences can also be broken down until we have a sequence of 1 point whose Fourier transform is that point. This approach can be satisfactorily programmed to yield an F.F.T. A complete Fourier transform can be obtained quite quickly and simply through simple additions and multiplications. The sequence of operations for 8 data points is shown in figure III.1.

Appendix IV

Revision of matrix algebra

Consider the simultaneous equations

$$\begin{aligned}
Y_1 &= a_{11}x_1 + a_{12}x_2 + a_{13}x_3 \ldots a_{1n}x_n \\
Y_2 &= . \qquad a_{22}x_2 \quad . \qquad \ldots a_{2n}x_n \\
Y_3 &= . \qquad . \qquad a_{33}x_3 \ldots a_{3n}x_n \quad . \\
. \quad & . \quad . \quad . \quad \ldots . \\
Y_n &= a_{n1}x_1 + a_{n2}x_2 + a_{n3}x_3 \ldots a_{nn}x_n
\end{aligned}$$

A shorthand notation is achieved by the use of matrices as follows

$$(Y_1, Y_2, \ldots, Y_n) = \begin{pmatrix} a_{11} & . & . & \ldots & a_{1n} \\ . & a_{22} & . & \ldots & a_{2n} \\ . & . & a_{33} & \ldots & a_{3n} \\ . & . & . & \ldots & . \\ a_{n1} & . & . & \ldots & a_{nn} \end{pmatrix} \begin{pmatrix} x_1 \\ x_2 \\ x_3 \\ . \\ x_n \end{pmatrix}$$

where $(Y_1, Y_2, \ldots, Y_n)$ is a $1 \times n$ row matrix or vector,

$$\begin{pmatrix} a_{11} & . & . & \ldots & a_{1n} \\ . & a_{22} & . & \ldots & a_{2n} \\ . & . & a_{33} & \ldots & a_{3n} \\ . & . & . & \ldots & . \\ a_{n1} & . & . & \ldots & a_{nn} \end{pmatrix} \text{ and } \begin{pmatrix} x_1 \\ x_2 \\ x_3 \\ . \\ x_n \end{pmatrix}$$

are, respectively, an $n \times n$ matrix, and an $n \times 1$ column matrix or vector.

Multiplication of vectors

Consider

$$\phi = ax + by + cz + dw \ ,$$

which we may write as the product of the variables in the form of a vector (x, y, z, w) and the coefficients also in the form of a vector as follows:

$$\phi = (a, b, c, d) \begin{pmatrix} x \\ y \\ z \\ w \end{pmatrix} .$$

This example shows the basic rule involved in multiplying matrices.

A matrix is simply a rectangular array of numbers, coefficients, or variables, each row and column of which can be regarded as a vector.

If

$$A = \begin{pmatrix} a & b \\ c & d \end{pmatrix}, \quad B = \begin{pmatrix} x & y \\ z & w \end{pmatrix},$$

then applying and extending the rule for multiplication we have

$$AB = \begin{pmatrix} a & b \\ c & d \end{pmatrix} \times \begin{pmatrix} x & y \\ z & w \end{pmatrix}$$
$$= \begin{pmatrix} ax+bz, & ay+bw \\ cx+dz, & cy+bw \end{pmatrix}$$

whilst

$$BA = \begin{pmatrix} ax+cy, & cx+dy \\ az+cw, & bz+dw \end{pmatrix}.$$

Note that, in general, $AB \neq BA$. Also in order that two matrices can be multiplied it is necessary that the number of columns of the first matrix is the same as the number of rows of the second matrix.

Transpose of a matrix

If A is a matrix with M rows and N columns, another matrix may be obtained from it by changing rows into columns and columns into rows. Such a new matrix is called the transpose of A and written A'.

For example if

$$A = \begin{pmatrix} a & b \\ c & d \\ e & f \end{pmatrix}$$

then

$$A' = \begin{pmatrix} a & c & e \\ b & d & f \end{pmatrix}.$$

This notation can help in the summarising of linear equations. For example the linear set of equations at the beginning of this appendix can be written down with this notation as

$$Y = AX' ,$$

where Y and X' are row vectors and A is a square matrix.

Appendix V

Heuristic proof of the Wiener–Khintchine relationship

$$S_{xx}(\omega) = \int_{-\infty}^{\infty} R_{xx}(\tau)\mathrm{e}^{-\mathrm{i}\omega\tau}\,\mathrm{d}\tau$$

$$R_{xx}(\tau) = \frac{1}{2\pi}\int_{-\infty}^{\infty} S_{xx}(\omega)\mathrm{e}^{-\mathrm{i}\omega\tau}\,\mathrm{d}\omega\ .$$

Let us define as before the autocorrelation function as

$$R_{xx}(\tau) = \lim_{T\to\infty}\int_{-T}^{T} x(t)x(t+\tau)\,\mathrm{d}t\ .$$

The Fourier transform of the autocorrelation function is given by

$$\text{F.T.}[R_{xx}(\tau)] = \int_{-\infty}^{\infty} R_{xx}(\tau)\mathrm{e}^{-\mathrm{i}\omega\tau}\,\mathrm{d}\tau$$

$$= \int_{-\infty}^{\infty} \lim_{T\to\infty}\frac{1}{2T}\int_{-T}^{T} x(t)x(t+\tau)\mathrm{e}^{-\mathrm{i}\omega\tau}\,\mathrm{d}t\,\mathrm{d}\tau\ .$$

If we now multiply the integrand by $\mathrm{e}^{-\mathrm{i}\omega t}\,\mathrm{e}^{\mathrm{i}\omega t}$, which is equal to unity and hence leaves the integrand unchanged, we obtain after some rearranging

$$\text{F.T.}[R_{xx}(\tau)] = \int_{-\infty}^{\infty} \lim_{T\to\infty}\frac{1}{2T}\int_{-T}^{T} x(t)x(t+\tau)\mathrm{e}^{\mathrm{i}\omega t}\mathrm{e}^{-\mathrm{i}\omega(t+\tau)}\,\mathrm{d}t\,\mathrm{d}\tau\ .$$

We now recognise that the integrand can be split into two parts which are

$$x(t+\tau)\mathrm{e}^{-\mathrm{i}\omega(t+\tau)} \qquad \text{and} \qquad x(t)\mathrm{e}^{\mathrm{i}\omega t}\ .$$

Both these are Fourier transform components, the second being the complex conjugate of the first. The expression then becomes

$$\text{F.T.}[R_{xx}(\tau)] = \lim_{T\to\infty}\frac{1}{2T}X_T(\omega)X_T^*(\omega)\ ,$$

where $X_T(\omega)$ is the Fourier transform of a truncated form of $x(t)$. The expression of the right hand side is by definition the power spectral density. Hence

$$\text{F.T.}[R_{xx}(\tau)] = S_{xx}(\omega)\ .$$

Index